EXPLORATION WITH A COMPUTER

GEOSCIENCE DATA ANALYSIS APPLICATIONS

COMPUTER METHODS IN THE GEOSCIENCES

Daniel F. Merriam, Series Editor

Previously published by Van Nostrand Reinhold Co. Inc.*

Computer Applications in Petroleum Geology
Joseph E. Robinson
Graphic Display of Two- and Three-Dimensional Markov
Computer Models in Geology
Cunshan Lin and John W. Harbaugh
Image Processing of Geological Data
Andrea G. Fabbri
Contouring Geologic Surfaces with the Computer
Thomas A. Jones, David E. Hamilton
and Carlton R. Johnson
Exploration-Geochemical Data Analysis with the IBM PC
George S. Koch Jr. (with programs on diskettes)
Geostatistics and Petroleum Geology
M. E. Hohn
Simulating Clastic Sedimentation
Daniel M. Tetzlaff and John W. Harbaugh

* Orders to: Van Nostrand Reinhold Co. Inc, 7625, Empire Drive, Florence, KY 41042, USA

Published by Pergamon

KOCH
Geological Problem Solving with Lotus 1-2-3 for Exploration and Mining Geology

Related Pergamon Publications

Book

HANLEY & MERRIAM (Editors)
Microcomputer Applications in Geology II

Journals

Computers & Geosciences

Computer Languages

Information Processing & Management

International Journal of Rock Mechanics and Mining Sciences (& Geomechanics Abstracts)

Minerals and Engineering

Full details of all Pergamon publications/free specimen copy of any Pergamon journal available on request from your nearest Pergamon office.

EXPLORATION WITH A COMPUTER

GEOSCIENCE DATA ANALYSIS APPLICATIONS

WILLIAM R. GREEN Placer Dome Inc., Vancouver B.C.

PERGAMON PRESS

Member of Maxwell Macmillan Pergamon Publishing Corporation

OXFORD · NEW YORK · BEIJING · FRANKFURT
SÃO PAULO · SYDNEY · TOKYO · TORONTO

U.K.	Pergamon Press plc, Headington Hill Hall, Oxford OX3 0BW, England
U.S.A.	Pergamon Press, Inc., Maxwell House, Fairview Park, Elmsford, New York 10523, U.S.A.
PEOPLE'S REPUBLIC OF CHINA	Pergamon Press, Room 4037, Qianmen Hotel, Beijing, People's Republic of China
FEDERAL REPUBLIC OF GERMANY	Pergamon Press GmbH, Hammerweg 6, D-6242 Kronberg, Federal Republic of Germany
BRAZIL	Pergamon Editora Ltda, Rua Eça de Queiros, 346, CEP 04011, Paraiso, São Paulo, Brazil
AUSTRALIA	Pergamon Press Australia Pty Ltd., P.O. Box 544, Potts Point, N.S.W. 2011, Australia
JAPAN	Pergamon Press, 5th Floor, Matsuoka Central Building, 1-7-1 Nishishinjuku, Shinjuku-ku, Tokyo 160, Japan
CANADA	Pergamon Press Canada Ltd., Suite No. 271, 253 College Street, Toronto, Ontario, Canada M5T 1R5

First edition 1991

Library of Congress Cataloging-in-Publication-Data
Green, William R.
Exploration with a computer: geoscience data analysis applications, c1990.
p. cm. -- (Computer methods in the geosciences)
Includes bibliographical references and index.
1. Prospecting--Geophysical methods--Data processing.
I. Title. II. Series.
TN269.G732 1990   622'.15'0285--dc20   90-43891

British Library Cataloguing in Publication Data
Green, William R.
Exploration with a computer: geoscience data analysis applications.
1. Geochemical analysis applications of computer systems
I. Title II. Series
551.9
ISBN 0-08-040264-X

Printed in Great Britain by BPCC Wheatons Ltd, Exeter

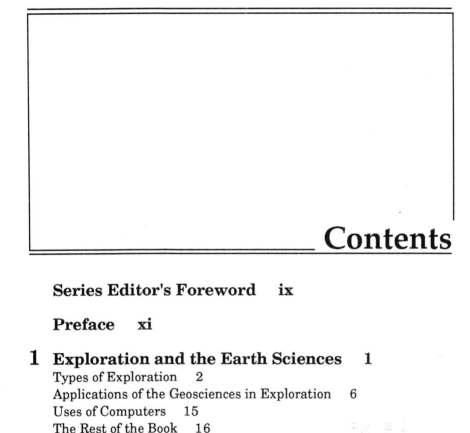

Contents

Series Editor's Foreword

Exploration with a Computer is a book on how to look for things—mineral deposits, to be specific. Bill Green focuses on the methods rather than the applications in this concise, well-written exposé on geoscience data analysis applications. As he notes in his preface, the mathematical details are outlined only briefly because they are covered in other books. He notes also that software is available and many applications have been described in the literature, so the problem remaining is how to use exploration methods effectively. And, that is what his book is all about.

Green groups the use of computers in exploration into three catagories: data entry, analysis, and interpretation. After a general introduction, he discusses aspects of data entry and requirements for computer analysis of all types of exploratory data and then, chapter by chapter, data analysis of geochemical, geophysical, remote sensing and image analysis, and drillhole data. Another chapter is concerned with problems of resource estimation and the final chapter is on other applications of computers, including preparation of reports, spread-sheets, communications, GIS, ES/AI, and exploration decision making.

One appendix gives guidelines for effective computer analysis, a second gives information in selecting a software system, and a third lists organizations, pertinent journals, and general references and available software by subject. All in all, the book is a complete package for "computerizing" the reader.

Because computers have taken over most of the manual methods both in the lab and in the field, it is important that the explorationist keep up with these fast developing approaches to stay current. Green brings to the reader discussions of these subjects from the point of view of an end user.

This contribution is the ninth in the series of Computer Methods in the Geosciences first published by Hutchinson Ross Publishing Co., then by Van Nostrand Reinhold, and now by Pergamon Press. There are several other books in this series that would be of interest to the reader of this book: *Computer Applications in Petroleum Geology* by Joe Robinson; *Exploration–Geochemical Data Analysis with the IBM PC* and *Geological Problem Solving with Lotus 1-2-3 for Exploration and Mining Geology* both by George Koch; and *Geostatistics and Petroleum Geology* by Mike Hohn. This book by Bill Green is a handsome addition to the series and will prove invaluable to the explorationist.

D. F. MERRIAM

Preface

This book is a guide to the uses of computers in exploration, where *exploration* is defined as the application of scientific methods to discover unknown features of the Earth's surface and crust.   This type of exploration includes the search for new mineral and petroleum resources, as well as such related problems as identifying geologic hazards and locating sources of pollution.   Some of the techniques have a long history of computer application (e.g., seismic exploration for oil), while others have only recently included use of the computer to replace manual methods (e.g. geologic logging of drill-core samples).

The work described herein is a follow-up to an earlier book, *Computer-Aided Data Analysis*, in which I reviewed many of the basic features of computers and their uses in fields where numerical and spatial analysis are important. *Exploration with a Computer* is more specific to the particular procedures involved in exploration: geochemical sampling, geophysical surveys, remote sensing, and direct sampling of the crust with drillholes.

The emphasis of the book is on the practical aspects of using computers in the earth sciences.   The principles of the various exploration techniques are given only in outline, to permit concentration on the problems of getting information into a computer-acceptable form, running programs in an efficient manner, maintaining flexibility in choosing analytical methods, and so on. Software to do all of the tasks related to exploration data analysis is readily available, so mathematical details have not been included. In the interests of clarity, specific references are not included in the text, but are listed and summarized in the Bibliography. Additional sources of information are given in Appendix C.

Because there is a strong trend toward technical staffs doing their own computer work (rather than relying on outside consultants or computer specialists in their organization), the book takes the point of view of an end user. Developers of computer software for the earth sciences should also find the book of value, as it stresses the features that make software practical and effective for the users.

Specific programs are not discussed; rather, the focus is on how to use the general capabilities of data analysis and graphical display. Many existing programs can provide these functions. The key to becoming an effective computer user is to learn by doing. No book can substitute for the experience gained sitting at a terminal working with real data. The objective of this book is to provide some guidelines, and to outline what is practical. I hope the suggestions of alternative methods will prove useful.

ACKNOWLEDGMENTS

I wrote *Exploration with a Computer* to supplement my earlier book with guidelines for my associates on effective ways to use the computer systems at PDI (Placer Dome Inc. and its predecessor, Placer Development Ltd.). The continuing development of exploration software at PDI provided much of the practical slant of this book. Many people within PDI have used the software, and provided many helpful suggestions on improvements. Indirectly they are contributors to this book. Although the large number of people involved precludes naming everyone, several have been particularly good at finding the weaknesses in the programs. Peter Kowalczyk and Stewart Nimmo in Vancouver, Dick Lewis in Sydney, and Dennis Geasan in Reno deserve special mention and thanks.

A second stimulus for writing the book was a holiday in Italy in 1986 (for this I must thank my wife, Nunzia, who had the good sense to be born there). By coincidence, a NATO ASI conference on Statistical Treatments for Estimation of Mineral and Energy Resources was held in Italy at the same time. After some complicated interational phone calls to locate the meeting (in Lucca), we managed to take a side trip to spend three days there. Many of the leading figures in the field were taking part, including John Davis, George Koch, Daniel Merriam, and Frits Agterberg. The opportunity to talk to people from around the world, and to meet the names behind the standard references, was an inspiration and worth the trip in itself. The concept of writing a book specifically on exploration was nebulous in my mind before then, but the encouragement from others at Lucca solidified the idea. By the end of the meeting I already had a brief outline.

I would like to thank John Davis in particular for his enthusiasm about my first book. Dan Merriam suggested this series as a place to publish the work, and encouraged me to get an outline and proposal ready. He and Chuck Hutchinson later carried the proposal through to the publishing agreement. After some stops and starts on my part, and a change of publishers, here it is. I hope the finished product lives up to the initial expectations.

CHAPTER **1**

Exploration and
the Earth Sciences

This book is about *exploration*, in a fairly general context meaning the search for unknown features of the Earth. In this sense it follows the historical meaning of undertaking voyages to previously unknown parts of the world. More modern types of exploration into the microcosmic and macrocosmic worlds are excluded (e.g., high-energy physics and molecular biology for the former, and astronomy, space exploration, and related fields for the latter). We will consider regions to be explored on scales related to everyday experience, that is, of a size we can at least contemplate walking over or viewing with the naked eye.

The process of exploration is certainly as old as humankind, although its nature has changed through the centuries. The geographical aspect of charting unknown territories is a thing of the past (at least on Earth). Today exploration is largely the search for undiscovered resources under the Earth's surface, or for characteristics of the surface not detectable by the human senses. Here *resource* should be considered as a broad category, defined as any hidden features of interest to scientists, including things not normally considered resources, for example archeological objects and polluted areas.

Searching for resources such as minerals has also been going on for a long time, of course. The history of past successes is in large part responsible for the difficult task facing modern explorers: All of the easy discoveries already have been made. In general we can say that the remaining resources are well-hidden. The result is that exploration is becoming an increasingly complex operation, relying on sophisticated technology and complex mathematical calculations.

A key to exploration success is the ability to detect subtle patterns in scientific observations of the Earth's surface and near surface. While in the past detection came from visual examination and application of experience and judgment, modern methods often involve phenomena completely undetectable by the human senses. It is in these areas that the computer takes a primary role in detecting exploration targets, providing the ability to perform the complex calculations needed to analyze exploration data.

Computers do much more for exploration than analyze unusual types of information. As in all facets of modern society, they are increasingly taking over manual and repetitive tasks (such as the drawing of maps). For earth scientists involved in exploration, the result is that they can spend more time in developing an understanding of geologic processes. In the long run this should lead to more discoveries of new resources, although as we have already noted the power of computer methods is at least partly offset by the difficulty of finding the well-hidden remaining targets.

TYPES OF EXPLORATION

Searching for Mineral Resources

Perhaps the oldest application of resource exploration is the search for concentrations of mineral (especially metallic) ores, which has a history dating back thousands of years. What comprises *ore* is not always clearly known, since its definition is tied to economics: If a body of mineralized rock can be extracted from the ground at a profit, it is ore; otherwise it is not. Because changing metal prices and other factors can change the category, the task of mineral exploration is to locate mineralized bodies that have a high probability of becoming ore, given the economic conditions of the day. (That is, whether the discovery is truly ore is not known until it is being mined, which follows after the exploration is complete.)

The first step in mineral exploration is to define areas likely to contain deposits of a particular economic mineral (or at least have a higher than average probability of such an occurrence). Then such areas are examined in detail, using methods to be outlined in the next section. Given a measure of luck, and a thorough application of geologic principles and experience, a previously unknown concentration of potential ore may be found.

There is more to mineral exploration than finding a deposit of mineralized rock, however. Before a deposit can be mined successfully, its characteristics must be determined in detail (size, orientation, concentrations of metals or "grade," nature of the host rock, and a variety of other parameters must be defined). In addition to defining the resource, the exploration process must also anticipate the problems of its development. Thus data pertinent to the engineering design of a mine should be acquired at the same time, if possible.

Finding New Sources of Oil

A similar problem to mineral exploration is the search for new fields of oil and natural gas. Although petroleum exploration has a shorter history, it is of even greater importance to the present world economy, which is so heavily dependent on oil and gas as primary sources of energy.

The basic strategy of first identifying promising areas to be examined in detail also applies to oil exploration. There is a conceptual difference in defining an area, however: Oil and gas may be extracted economically from greater depths than minerals, and thus *area* implies a volume in a three-dimensional sense. Mines also are concerned with volumes, of course, but generally are restricted to situations easily accessible from the surface. Deeper mines are developed by continued exploration from the underground workings during the mining phase, rather than from the surface.

There are a number of oil fields in the world located under the same part of the Earth's surface, but contained in completely different geologic formations. In some cases, deeper sources of oil were long overlooked, as it is easy to assume an area is extensively explored by simply looking at the density of drilled wells, without checking the number of wells that reached particular formations at depth.

Due to the deep burial of oil and gas, there may be little direct surface expression. As a result, oil exploration is more dependent on indirect techniques such as seismic profiling, and thus more dependent on computers to aid in data analysis. Both mineral and petroleum exploration make extensive use of drilling to obtain samples of the resource region. The greater depths typical of oil wells mean that drilling is more expensive in this situation. (Petroleum exploration in general is many times more expensive than mineral exploration, although exploration costs may be offset in part by lower development costs when a field is put into production.)

Hydrological Studies

One of the most critical resources may be taken for granted, but may be dangerously short in supply. Water is a renewable resource, but must be studied carefully to determine maximum extraction rates. In addition, sources of pure water are increasingly hard to find, as rising populations and industrial requirements tend to contaminate existing supplies at the same time as demand increases.

Finding water supplies has many similarities to mineral and petroleum exploration, since it is a problem of locating favorable geologic situations (aquifers). Like oil, fluid movements under the ground must be charted as well, since both source and reservoir regions are involved. Estimation of quantities is also an important factor, which might be the volume of an aquifer, or the winter snow pack that will serve as the source of summer water supplies.

The last example indicates a significant difference in hydrology: The characteristics of theresource change with time. These changes must be determined before effective exploitation strategies can be developed. The only way this determination can be achieved is by doing repeated surveys over a long period, due to natural climatic fluctuations. This is especially important when large watersheds are involved. The danger in neglecting long-term variations is illustrated by the Colorado River. Current use of its water runs the river virtually dry before it reaches the ocean. This is mainly the result of sharing agreements made in the early part of the century, that were based on flow rates observed for only a few years. These rates turned out to be considerably higher than the longer-term average.

Site Planning: Locating Geologic Hazards

Many of the techniques for identifying geologic structures and other properties of the near surface are also of importance in civil engineering. In this instance the objective is not to locate geologic features that are of value in themselves, but to avoid potential problems in construction of buildings, bridges, pipelines, and other structures. In other words, exploration is aimed at reducing the chances of financial losses, rather than making gains.

The area of interest in most engineering applications is much smaller than for the examples discussed above. This does not imply that the task is easier (or cheaper), since the level of detail required (i.e., the spatial resolution) is correspondingly higher. For example, in planning an open-pit mine, it may be sufficient to describe the rock properties as averages within blocks 10 meters on a side. To adequately plan the foundation for a major building geologic maps may well have to accurately show features well under a meter in size.

Another difference in exploration for site planning is that there is generally not an initial phase of regional study leading to detailed follow-up, as is the norm in exploring for minerals, petroleum, and water supplies. The sites of engineering projects are usually chosen on criteria totally unrelated to the geologic setting. An exploration stage may be required to select the safest (or cheapest) of competing sites, or to define critical construction parameters (such as the depth of foundations). Exploration is thus an adjunct to the planning process, and not usually a primary factor in deciding whether a project will proceed. It may have profound effects on the construction phase, however.

Geotechnical studies may have a large hydrological component, because groundwater is a major factor in soil stability. Environmental impact of engineering projects is also a major concern, especially for chemical plants and similar operations in which hazardous or toxic materials may be introduced into the ground.

Environmental Monitoring

This procedure is closely related to some aspects of site planning, but is an ongoing process. Like hydrological studies, the data also have a time dependence, and part of the exploration process is to find areas where trace-metal concentrations or other contaminants are changing with time. Tracing contaminants to an unknown source may be needed, along with some geochemical detective work to match the pollutants to one of several possible sources.

Investigation of Archeological Sites

The examination of archeological sites using modern exploration techniques is like the engineering problem, in that high spatial resolution is needed. It also has similarities to the mineral and petroleum exploration problems, in that the main objective may be to locate unknown objects of great value. The objective may also be more like the engineering situation, when the problem is to determine how to excavate a site most effectively and efficiently.

In some cases, the pattern of regional study leading to detailed work also applies here. For example, many prehistoric cultures built numerous burial mounds that are heavily obscured by more recent societies. Regional methods may be used to detect probable locations of former mounds, which are then subjected to the highly detailed techniques appropriate to the local sites.

APPLICATIONS OF THE GEOSCIENCES IN EXPLORATION

Exploration in the sense of this book relies on a thorough knowledge of the earth sciences. *Geoscience* is the study of the Earth and the processes that shape it. It involves the classical physical sciences (physics, chemistry, mathematics), as we might expect, but also many of the life sciences, since many geologic processes are strongly influenced by past and present living organisms.

Traditionally geology meant the study of the Earth in all its forms. With scientific specialization being the norm here, as in other fields, it is now usual to consider geophysics, geochemistry, geobotany and others as separate (although obviously related) sciences. Geology in exploration then may be taken to mean the direct study of Earth's features (e.g., visual examination of rock or soil samples), or application of basic principles of earth-forming processes (e.g. development of structures). The other earth sciences then imply a more indirect approach to determining the Earth's characteristics. Obviously there can be real difficulty in classifying some areas of the study of the Earth: Is a theory of mountain building pure geology or geophysics, for example.

Exploration is almost always an interdisciplinary process, involving the use of several sciences. In the following discussions of techniques in the specific fields, it should be borne in mind that the results of one procedure may be used together with other methods in deciding what to do next. In addition, combined interpretation of different types of data may well resolve ambiguities in all of the individual surveys. Many of the examples can be considered at least partly engineering problems, in that the ultimate objective is not simply to obtain new knowledge. The application of geosciences may be aimed at exploitation of a resource, or it may be part of an engineering design.

We will start with geology, because it is the parent to all of the others. The other methods will follow in the sequence typically (although certainly not always) followed in a long-term exploration program.

Geological Exploration

In many ways, this term applies to all aspects of exploration, from initial regional overviews to detailed examination of small mineralized zones, structural features, and so on. Application of sound geological principles is essential in all phases of exploration. The results of the specialized survey techniques to be discussed below can be fully understood only in the context of an acceptable geological model.

Prospecting and Mapping

What first comes to mind when we think of geologists embarking on a search for new resources is the stage of going to "the field" to study an area. This is by far the oldest method of searching for resources. In the simplest form, a geologist examines an area on foot, noting the dominant types of rocks, structural form of the surface, and presence (or absence) of particular minerals known to be related to metallic ores. In many parts of the world, this task can be difficult, since most of the surface is covered by transported soils and vegetation. In other words, "outcrops" of the underlying rocks are essential to the process.

These observations in the field are commonly incorporated into a geologic map, so that spatial trends or associations can be detected. With the aid of recent technological advances (such as stereoscopic aerial photographs and satellite images), it is now possible to do much of this work on a regional scale before going to the field to look at the rocks.

In the earlier days of the petroleum industry, this approach was also instrumental in finding major oil fields, by locating surface expressions of deeper structures and in some cases traces of hydrocarbons. Modern oil exploration usually is aimed at deeper structures, so that surface geological examination does not play such a key role. (As mentioned at the outset, the easy discoveries have for the most part already been made.)

The prospecting approach to mineral exploration is to look for rocks that might contain economic quantities of ore minerals (without necessarily combining this with a general geologic mapping program). There are many historical examples where a large measure of luck is involved in making significant finds, generally by people with no formal training in geoscience. Still, knowledge and experience in geology greatly improve the odds in prospecting, and reduce the chances of overlooking potentially valuable mineral occurrences.

Although many mineral ores are visually distinctive, it is difficult to estimate their metal content (grade) without doing a chemical analysis. For this reason, a large part of the prospecting job is to collect samples of potential ore-bearing rocks, and to map their extent. After assays are returned from a laboratory, it is then possible to make at least a rough estimate of the potential value of any discovery. (This simplification is obviously extreme, as we shall see later in describing the many problems involved in mapping and sampling.)

Modeling and Interpretation

In modern times, the application of geology to exploration goes far beyond the traditional prospecting approach. Because most deposits are small (relative to the land area of the Earth), it is impossible to sample the entire surface. A major role for the geologist then is to apply modern concepts of how ore deposits are formed, to identify the most likely areas for particular minerals to be located. This procedure is becoming increasingly dependent on the principles of plate tectonics, which since the late 1960s has led to a more complete understanding of how the features of the crust are created and modified.

This initial application of geology is also important in petroleum exploration. Once again, the concepts of plate tectonics provide guidelines for defining areas likely to contain oil or natural gas (primarily areas that at some point in geologic history had large rates of sedimentation combined with heat sources to convert organic material into hydrocarbons). The geologist also must understand the mechanisms by which oil can migrate through the rocks to locate the regions within the sedimentary basin that likely trapped the oil in a recoverable reservoir.

Even after a deposit has been identified, this aspect of geology remains of prime importance. Developing a conceptual picture of how the potential ore minerals were deposited is a key to the success of the detailed exploration needed to determine the true size and economic value of the deposit. Such geological models are revised as new exploration techniques or extensions into unexplored areas provide new data to add to the total picture.

In geotechnical applications, detailed models of structural geology are the main objective of site investigations. In addition to the structural configuration, the model requires parameters (such as fracture density and compressive strength) needed to determine the ability of the site to support the proposed structure.

Remote Sensing

Although this technique is the newest and most technologically sophisticated of common exploration techniques, it is usually one of the first steps, and can be done prior to any field work. Remote sensing is any method of surveying an area from a distance. In the most general sense, it includes simple photography, and most of the geophysical methods to be described later. In common use, the term is reserved for techniques outside such well-established fields, in particular for methods using various types of electromagnetic energy (infrared radiation, radar, etc.). These methods may be considered passive (simply recording naturally existing energy) or active (where the return of energy from an artificial transmission is recorded, as in synthetic aperture radar [SAR]).

Most methods involve recording the energy in various wavelength bands, for a regular array of cells covering the survey region. Each cell is termed a *pixel* (from picture element, since the array of cells can be displayed in a two-dimensional image just like a television picture). The related field of image processing provides the computer hardware and software to convert the arrays of remotely sensed data into pictures that can be visually interpreted.

Remote sensing is often taken to mean only the application of satellite sensors to map the Earth's surface. Aircraft surveys employ the same types of instruments, however, and should be considered as equally important. This consideration is especially true when the resolution requirements of geologic mapping are taken into account. In addition, aircraft surveys can generally record more types of data, because the data can be recorded directly on magnetic tape, and do not have to be transmitted through a relay system.

Geochemical Sampling

Geochemical exploration can be considered a direct method like taking rock samples. The basic procedure is to collect samples of soils, sediments in streams, or similar materials and analyze them for trace contents of metallic elements, hydrocarbons, toxic compounds, and so forth. It also has a strong indirect aspect, in that erosion and/or hydrological processes can carry these materials far from their source.

In mineral exploration, another indirect aspect is that the metal of interest may not be detected itself, but may be indicated by the presence of other elements that are commonly associated with it.

Because of these characteristics, geochemical surveys are usually part of the earlier stages of exploration. They are aimed primarily at indicating areas likely to contain mineralized deposits or other targets, rather than to directly detect or delineate them.

In regional scale mineral exploration, stream sediment sampling is a widely used technique. Fairly complete coverage of a large area can be obtained by careful planning of sample sites to correspond to the drainage network of the region. After analysis of samples, results are plotted on maps, and samples with anomalous concentrations of various elements are identified to mark areas for follow-up sampling. The follow-up might be another round of stream sampling, going into finer branches of the drainage system, since the source of anomalous material may be anywhere in the drainage catchment area.

On a more local scale, soil samples may be collected on a regular grid over areas showing promise in regional surveys. After collecting the samples, the procedure is essentially the same: plotting analytical results on maps to identify localized anomalous zones. If any of these are sufficiently promising, the next stage probably would be direct sampling of the subsurface by drilling, which we will discuss later.

Geochemical methods are also applied in oil exploration, although not as routinely as for minerals. Here the objective is to find traces of hydrocarbons in surface rocks and soils, as indicators of deeper pools of oil or natural gas.

Geochemical analyses are also important in environmental problems, such as determining how far specific pollutants spread from their point of origin. Similarly, trace-element geochemical signatures may be useful in tracing flows in aquifers, and checking for ultimate sources of groundwater.

Geophysical Methods

As a general definition, geophysics is the application of the principles of physics to the study of the Earth. In exploration, geophysics usually means observation of various types of transmitted energy to determine the characteristics of the region below the surface of the Earth, which obviously is not easily accessible for geological examination.

A variety of exploration methods which fall into this category. Geophysicists have been inventive in devising techniques to exploit almost every type of energy source, and to apply them to many different geological problems. There are many profound differences in the various types of surveys, but also strong similarities (due to the similar nature of the problems to be solved).

Like geochemistry, geophysical surveys may be done on a regional basis, with rapid but low-resolution coverage of large areas. When interesting anomalies are found, more detailed local surveys follow (frequently involving different exploration methods). Anomalies outlined in local surveys are tested by drilling to provide direct sampling.

By far the most extensively used geophysical method is seismic reflection profiling, which is the fundamental technique for oil exploration. The procedure involves generating artificial seismic waves at the Earth's surface, and recording waves reflected back from internal structures in the crust at a number of points on the surface (along a line or on a two-dimensional grid). By taking many "shots" at different locations, and moving the detectors as well, it is possible to build a detailed three-dimensional view of the structural layers that may form traps for oil and natural gas.

High-resolution seismic surveys are also used in site investigations. The procedures are essentially the same as in petroleum exploration, except that the spacing of sensors is smaller, and the instrumentation is designed to record higher frequencies.

Ground-probing radar systems provide another way to map near-surface structures in great detail. The operational procedures are again similar to conventional seismic reflection. The different characteristics of the energy allow this technique to be used in situations where seismic fails, for example, testing for defects in concrete road surfaces.

Less detailed structural information may also be obtained from gravity and magnetic surveys (taken together as "potential field" methods, from the mathematical formulation used to define their physical properties). These methods are passive, and they simply require recording the gravity or magnetic fields at a number of stations on the surface. For regional surveys, the readings can also be taken from aircraft, thus making it possible to survey a large area quickly and inexpensively. This aspect is a major factor in using these methods in the early stages of oil exploration, since seismic mapping is too expensive to apply on a regional basis. Potential field methods are also useful complements to seismic methods on local surveys, since they provide other information about the subsurface structures (rock density and magnetic characteristics, for example).

There are several methods that use electrical or electromagnetic energy propagation. As a general rule, these are used mostly in mineral exploration, because rock conductivity and similar parameters are very useful in identifying mineralized rocks. They usually are not as diagnostic for studying possible oil fields, which are hosted in sedimentary rocks of similar electromagnetic characteristics.

These techniques include both active and passive types, although the terms here have a slightly different context. Active methods such as induced polarization involve generating electric currents (or electromagnetic waves) and recording the transmitted or stimulated results. Passive here means that an artificial source is not required, although the methods rely on measuring the Earth's response to some other energy source. For example, magnetotellurics measures electric currents induced by natural variations in the Earth's magnetic field. Very low frequency (VLF) measures the Earth's response to long wavelength electromagnetic waves used as carriers for worldwide military communications from fixed land stations to submarines.

The basic design of a geophysical survey is similar to geochemistry, in that the objective is to take samples (here meaning geophysical measurements) over a designated area, and examine the results in map form to locate anomalous regions for more detailed study or to find drilling targets. There are some characteristics that are unique to geophysics, however. First, results usually require special calculations to put raw readings into understandable form. With most geophysical instruments now being integrated with microcomputers, such data reduction is normally done in the field, so that results are available for interpretation immediately. This is a major contrast to geochemical surveys, which usually require samples to be sent to a laboratory for analysis, often resulting in delays of weeks between collecting samples and knowing their mineral content.

However, results of geophysical surveys are not always obtained quickly. The significance of the recorded data may come to light only after detailed computer modeling to determine geologic parameters consistent with the observations. Many types of geophysical data are notoriously ambiguous (in that many different geologic models satisfy the same data), so it also may be necessary to work simultaneously with several different types of data to produce a unique solution.

Drilling

The final stage in exploration for resources below the Earth's surface is to take direct samples of the prospective region by drilling into it. In mineral exploration, the objective is to extract "core" samples of the rocks and measure their metal content. In oil exploration, the drillhole (well) tests favorable formations for the presence of hydrocarbons, and provides a path for extracting fluids from the buried geologic structures. In either case, a geologist normally monitors the progress of the drilling in the field, and is responsible for determining when (and if) the target structure has been reached, selecting zones to be sampled, and making other exploration decisions.

Drilling is similar to other exploration methods in that samples representative of discrete points in space are analyzed, and the results are incorporated into a spatial interpretation. It has unique problems, however, in the three-dimensional orientation of the samples. Special surveying techniques are required to determine the track of the drillhole in the crust. Holes may be intentionally deviated to reach targets not accessible from directly above, or to hit the target at an optimum angle. In addition, it is not always possible to completely control the path of the drill as it encounters rocks of different densities. As a result, a major part of analyzing and interpreting data from drillholes is to ensure that the positions of the samples are calculated correctly.

In addition to obtaining laboratory analyses of drillhole samples, the geologist normally logs the hole, noting changes in geologic characteristics with depth. This process is mostly manual, requiring careful examination of the core (or drill cuttings). In oil wells, logging also refers to geophysical surveys taken by recording the response of various "tools" lowered into the hole (gravity meters, gamma-ray sources, etc). In this case, logging is a more automated process, normally performed by specialists. The exploration geologist of course must understand the results and know how to integrate them into an interpretation of all available data.

While the prime objective of drilling is to make a direct hit on an oil zone or mineralized deposit, barren or dry holes are still of value in building a detailed geologic model of the survey region. Geophysical measurements in such holes may indicate new targets. Traces of minerals or hydrocarbons may encourage further exploration, although not constituting a discovery in themselves. In these situations, the new drilling data may be instrumental in choosing new drilling locations. Of course, there is always the chance that negative results will lead to a decision to abandon the prospect.

USES OF COMPUTERS

The remainder of this book examines how computers are used in various stages of exploration. As we shall see, some areas have a long history of computer use, while others only recently have adapted to computer methods. A brief outline follows here.

The uses of computers in exploration can be grouped roughly into three main categories. First is data entry or recording, in which the computer is used to store raw exploration data. Second comes data analysis, in which information is extracted from the data by mathematical and statistical operations. Finally there is interpretation of the data, in which the information obtained from the data is used to derive geologic models of the exploration target. In some fields computers are indispensable in interpretation because models are derived using complex numerical operations.

Data entry may be almost entirely automatic, as in many geophysical surveys. The instruments that measure the various geophysical parameters are integrated into specialized computer systems, and data are recorded directly into computer memory, or written onto magnetic tapes or disks. Depending on the software in the system, it also may be possible to add various supporting data from a keyboard (such as line numbers and locations). With more subjective data such as geologic logs, it is common to use a coding format suitable for typing into a computer. In some cases, portable computers may be used directly in the field, in place of writing on a form to be typed later.

Data analysis in exploration covers a wide variety of techniques. These may be statistical in nature, or essentially visual (contour maps, for example), or a combination of the two. Many of the methods can be done manually, provided the volume of data is not excessive, but generally can be done faster with a computer. The real advantage of using the computer is when many variations on the same type of analysis are required. To draw ten maps showing ten different geochemical analyses takes little more time than one map on a computer, since many of the basic parameters of the maps do not have to be changed. Manual drafting would of course take about the same amount of time for each additional map.

Although many of the data analysis methods are applicable to many different types of data, modeling and interpretation schemes tend to be specific to each type. For this reason, experience gained in one type of modeling is not necessarily helpful in studying another type. The graphical methods of analysis are often helpful, though, when the results of modeling are to be presented in visual form.

THE REST OF THE BOOK

In the following chapters, we will see how data entry, data analysis, and data interpretation are applied for various exploration methods. The discussion will focus on the method, rather than on its application (i.e. geophysical data analysis is a general topic, largely independent of whether the technique is applied in mineral or petroleum exploration, or to some other problem). Mathematical details will be outlined only briefly, since they are well covered in other books. In addition, software to do most applications is readily available: The challenge is learning how to use it effectively.

Chapter 2 reviews the requirements for computer analysis of all types of exploration data. It is followed by chapters on geochemical data analysis, geophysical data analysis, remote sensing and image analysis, and analysis of drillhole data. In Chapter 7, the problems of resource estimation are considered.

We might also consider a fourth category of computer use that does not fall into the natural sequence of acquisition, analysis and interpretation. Many of the clerical tasks required to manage a geoscience or engineering project now are done on computers. Most of these tasks are not restricted to the geoscience or engineering fields: budgeting, economic forecasting, report writing and other tasks are part of most business and scientific activities. These topics will also be discussed in the final chapter. In addition, some specialized applications such as decision making and large-scale data bases will be reviewed briefly.

Geologic computer applications are not discussed as a separate topic, even though geologic mapping and interpretation are considered to be the guiding framework for other types of exploration. The computer does not play as direct a role in geologic exploration (in the sense described above) as in its related disciplines, however.

This difference is a result of the subjective nature of geologic modeling: A healthy dose of experience and intuitive judgment is usually included in any interpretation.

Many of the techniques used in studying specific types of exploration data can be of value in geologic interpretation, especially in computer graphics and map plotting. The developing field of artificial intelligence (and within it expert systems) holds considerable promise for aiding in interpretation and modeling: In the future it is likely that computers will have more direct applications to geology. Applications of artificial intelligence will also be discussed in the final chapter.

Requirements for
Geoscience Data Analysis

As we saw in Chapter 1, there are many different exploration methods, each with its own characteristics. There are also many common features, which is a considerable help in using computers to aid in studying the data collected in various surveys. Many of the computer programs and procedures developed for one type of data may be directly applicable to other types, or may need only minor changes.

The general principles of computer data analysis will not be discussed in this book. They are covered in my earlier book, *Computer Aided Data Analysis*, which also reviews computer mapping and other applications of computer graphics in data analysis. These techniques now are nearly essential in any field involving spatial data analysis.

PREPARING DATA FOR COMPUTER ANALYSIS

The problems in acquiring correct data for analysis go beyond inherent uncertainties in measurement. The simple task of entering data into a computer can undo much careful work in acquiring and checking the data. Unfortunately, this step is often taken for granted. When combined with the tendency to consider measured coordinates as immune to errors, this tendency may lead to wasted effort in analyzing data that later prove to be invalid.

Exploration data may be grouped into three types. First are coordinates that describe the location of a sample. These are independent of the characteristics of the area of exploration interest. Second are numeric data associated with each sample (e.g., recorded geophysical fields or chemical assays), which are amenable to mathematical and statistical operations. Third are nonnumeric observations (e.g., the type of rocks outcropping in the vicinity of a stream geochemical sample). These generally have a relatively small number of allowed values, which can be used to divide the complete set of data into smaller groups. The behavior of the numeric variables may be different in each group.

The problem of getting accurate data into a computer is not confined to any single type, of course. Errors may be present in all types of data, and all phases of data entry require careful quality control to ensure that later analysis is based on valid information.

Data Entry

Many types of exploration data are collected and prepared for analysis entirely with computers. This method removes the possibility of errors introduced in manual transcription. It does not indicate that such data can be assumed to be error-free, however. There may be instrument malfunctions or noise on transmission lines to corrupt the readings. It is often necessary to manually add such ancillary data as line numbers or starting times, which obviously can be incorrect. In addition, there is the possibility of natural noise inherent in the data, even when all components of the data collection system function perfectly (e.g., doing a magnetic survey during a period of magnetic storms induced by solar activity).

When data are not recorded automatically, transcription from some other form is required, which usually involves entering numbers and text from paper records. In the past, the only available method was to type the information into a computer file from a keyboard. Scanning systems are often applied in this area now. This procedure requires hardware to represent a printed page as a digitized image, and software to detect patterns in this image corresponding to alphanumeric characters. Whether input is by typing or scanning, there is a considerable probability that transcription errors will be present. In scanning systems, a 95 percent accuracy rate is considered good, for example. Appropriate checking procedures must be implemented to detect and correct errors (see below).

Another source of trouble prior to data analysis is the data reduction calculations needed to convert the raw observations to interpretable form. For example, the readings from a gravity meter must be corrected for instrument drift, elevation effects, variation with latitude, the influence of surrounding terrain, and a variety of other perturbations before they can be related to density differences in the near surface. Similarly, in seismic profiling, the raw data consist of millions of samples of seismic wave amplitude for various spatial positions and at repeated time intervals after generating the waves. To create a seismic cross-section, these data must be adjusted to correct for different geometries of source and receiver, different near-surface geology, variations in seismic wave velocity in the crust, and natural decay of amplitude with length of the wave path. In addition, there is often a poor signal-to-noise ratio, meaning that the natural seismic noise has a large amplitude compared to the reflected waves. Sophisticated computer processing and averaging (stacking) of many readings is required to make the reflections visually detectable.

Data from more than one source are the rule rather than the exception in exploration. Thus a merging operation is needed as part of data preparation. There are a variety of ways merging may be done, most of which are relatively simple to implement. Successful merging demands some care in entering or checking the separate data sets: Common data must be marked by unique identifiers to guarantee that the correct match is made.

Detecting Input Errors

There is no magic answer to the problem of providing correct data. Data entry and verification will always be a time-consuming process, although it is possible to use computer methods to streamline the task.

With manually transcribed data, the traditional objective was to have the computer file match the original form; that is, to locate and correct all transcription errors. This procedure was usually done by a verification stage of data entry: typing all entries in twice, and checking any differences against the original. The other form of checking is to compare visually a computer listing with the source, which demands two people working together to be most effective.

Computer scanning for data-entry errors may be possible if the file has a well-defined format. For example, some fields may be assumed to contain strictly numeric values, which can be tested easily. Fields that contain arbitrary text are more difficult to scan, unless there is a set of allowable entries. For example, a historical data base may have country or state names as part of each record (or shorthand codes for them). By entering a list of valid codes into the computer, it is relatively simple to devise an automatic check. It may be done directly as the records are typed, if a special data-entry program is used. Alternately, the data may be entered by general purpose programs, and then passed through an error-checking routine specific to the particular type of file.

Manual checking against an original copy cannot be applied to recorded data, since by definition the original is already in computer form. (The vast quantities of most automatic recording would make manual checking impossible, anyway.) It may still be necessary to check that the data are in the correct format. There may be instrument malfunction in data acquisition. Noise in data transfer may introduce errors when data are transmitted from one computer to another without full error checking. The transfer method may be essentially a "blind send," and noise or delays on the transmission line may result in loss of data or insertion of extraneous characters.

These procedures are only the first level of error detection. When properly applied, they should ensure correct transcription and proper formatting of the data. They do not, however, guarantee that the recorded data are valid: It is entirely possible to have a computer file that is free of such errors, but that contains completely worthless data because of other problems (perhaps instrument malfunction, faulty collection procedures, etc.).

By analogy to grammar, these two facets of data validity may be termed syntax and logic. A sentence (or record in a computer file) may have proper spelling and form, but have no logical meaning when the context is considered. For example, consider a statement such as "George's hat ate supper in Paris." We need to go beyond rules of form to decide if the sentence is reasonable.

With exploration data, checking for logical validity means that the data do not violate any of the known characteristics for their particular type. Such tests then must define these general characteristics, and determine how well each set of data fits them, to identify samples that are potentially invalid. It may be fairly easy to catch gross errors by simple tests, but more subtle abnormalities are as likely to be legitimate anomalies as errors, making them more difficult to detect by simple scanning methods. For example, expected minimum and maximum values can be used on geochemical analyses (e.g., negative values are impossible), but a reading that is ten times greater than all of its neighbors is entirely possible.

The spatial nature of the exploration data (to be discussed in the next section) is exploited in more sophisticated methods. There is an underlying assumption in spatial data analysis that each sample bears at least some resemblance to its neighbors. In other words, the data are not distributed completely randomly in space. While this assumption may be a fairly large one, and not easily proven, it is essential to the exploration process itself. Recall that the goal of exploration is to define a region containing an economic resource. From past history we know that such regions are always localized (resources are not uniformly spread throughout the Earth). The definition of a region then depends on recognizing some type of characteristic that is generally different within the region when compared to the surrounding barren ground. Another way to consider the assumption of localization is that there would be no need for exploration if the distribution of resources was uniform.

The role of spatial analysis in error detection is essentially the same as in detecting significant anomalies (as we shall see in later chapters). In essence, the process is to compare each sample to its neighbors and identify those with sufficiently different values to be considered unusual. There are a great many computational methods that may be used; for now, it suffices to note that none of them should be difficult to use with properly written software.

To determine whether an unusual sample is a true anomaly or an error requires exercise of judgment and past experience. Generally there will be some rule of thumb as to how many adjacent samples must have anomalous values to be considered real. Such rules will differ depending on the amount of data, the expected accuracy of the measuring methods, the possible range of variability of the data, and a variety of other factors. They cannot always be easily integrated into a completely automatic method for screening data. The usual implementation of a scanning procedure using comparison to adjacent samples is to have the computer identify potential errors, but to leave the decision on whether to accept, delete, or correct the data values in question up to a skilled person.

CHARACTERISTICS OF SPATIAL DATA

The fundamental common feature of all methods of exploration is they collect spatial data, that is, information which is associated with some fixed location in space. In many cases the relationship is two-dimensional, usually being position on the Earth's surface. One-dimensional and three-dimensional situations occur frequently, as well; data of all three types may be involved in some surveys. For example, seismic surveys involve data taken at fixed increments of time, for different locations along a line (two different one-dimensional cases). Data from different lines are combined into maps representing structural surfaces below the surface (a two-dimensional view). The complete set of structures at various depths then comprises a three-dimensional geologic model.

Maps and Coordinate Systems

One of the fundamental problems in analyzing spatial data is to display data in a form that gives an appreciation of the relationships of individual samples. (Here *sample* is taken as a general term meaning a set of observations taken at a single location.) The usual solution to this problem is to produce some type of map, which in a general sense includes any two-dimensional graph, not just the traditional geographic meaning.

Because of the difficulties in producing three-dimensional models, or displaying a three-dimensional picture on a flat (i.e. two-dimensional) surface, maps are also the primary method for analysis of three-dimensional data. In this case, a series of maps are required to represent two-dimensional slices through a volume at selected values of one of the coordinates.

It is perhaps stretching a point to call graphs of one-dimensional data maps although such forms still have a two-dimensional aspect (e.g., a plot of magnetic field strength against time). The difference is that a data variable is used to control position along one axis of the plot, with the other axis representing a single coordinate. On conventional maps, data variables are plotted at positions controlled by two independent coordinates.

Following the mathematical convention, in this book I will refer to the horizontal axis of a map as the X coordinate, and the vertical axis as Y. For geographic maps with north at the top, X is a measure of "easting" (i.e. the numeric value of X increases to the east), and Y is equivalent to "northing." This convention sometimes causes confusion, since it is customary in mathematics to specify X before Y, while surveying practice has northing before easting.

When displaying data at various depths below the surface, the map is also a cross-section, with X representing distance along a surface trace, and Y representing the depth or elevation. Here too the mathematical convention may cause trouble, as it assumes that Y increases upward. If the vertical axis is depth below the surface, it decreases upward in the normal viewpoint with the surface at the top. Computer plotting programs that follow the mathematical convention may not allow the alternate system, so that depths must be expressed as negative numbers (or converted to elevations referred to sea level) before the cross-section can be drawn.

Defining the Map Frame

Before a map can be drawn, some basic parameters have to be defined. First the frame of the map must be set, which means defining the coordinates of the two axes, the limits of each axis, and the scale of the map. Although these tasks may seem trivial, in many situations the choices are not obvious. Each of these points will be examined in more detail.

The coordinates to be used in plotting maps are often determined at the outset of an exploration project. Where samples are taken at various places on the Earth's surface, their locations must be measured in reference to an appropriate survey system. When the exploration program is regional in extent, locations are referenced to the coordinates of topographic maps of the region. For many parts of the world, Universal Transverse Mercator (UTM) coordinates are used for standard maps. There may also be some other reference system for government maps and survey monuments, such as the various state plane systems used in many areas of the United States.

When a local area is to be studied in detail, a reference grid is usually set up for locating sample points. This grid allows fixed control points to be located at convenient positions in the working area, rather than using standard survey monuments that may be some distance away. The measured coordinates are smaller numbers that can easily be related to local landmarks, rather than unwieldy coordinates tied to a distant origin. For example, a local grid on a mineral claim may have an origin at the corner of the claim, so that the X-Y coordinates for any location can simply be converted into distance from the boundaries. If all points were referred to the UTM system, the corner coordinates might be, for example, $X = 493,612$ and $Y = 5,311,798$. It would take some mental gymnastics to relate a drillhole at $X = 495,820$ $Y = 5,312,604$ to its position in the claim (2,208 east and 806 north of the corner).

The selection of a geographic coordinate system for a project is not normally subject to frequent change. Several local grids may be involved at various stages of the project, perhaps to be converted eventually to a common system. Even so, it is likely that one system will be used continuously for a long time, and all locations referenced to it do not have to be adjusted. On occasion, coordinates need to transformed into another reference by translation and rotation. This function should be included in the data preparation programs used for general data analysis.

When dealing with frames of reference other than conventional maps, the X-Y coordinates may not be permanent or measured values.

In plotting cross-sections for various surface traces, the X axis is usually the distance along the line, which obviously is different for each cross-section. If the section is a projection onto an oblique plane, the Y coordinate probably will also be calculated independently for each section. This is a fairly common situation in mineral exploration, where mineralization is confined to some tabular body such as a fault zone or vein.

Importance of Accurate Location Data

The need to have accurate and correct location information cannot be overemphasized: The goal of exploration is after all to determine the precise position of some object (which may be an oil trap, gold-bearing vein, potential geologic hazard, or any of the other targets described in Chap. 1). This fact may be overlooked by the natural bias of thinking of coordinates as independent variables, as distinguished from other types of exploration data that are usually viewed in the mathematical sense of dependent variables.

Different methods of defining sample locations have different probabilities for generating errors. If samples are surveyed with modern electronic surveying instruments, the data are likely to largely error-free (within the precision level of the instrument). This fact applies both to the actual measurement and entry into the computer, since the latter step does not require typing in the values, but is done automatically by the computers that control the survey instruments.

Sample locations may be defined by more manual methods, which are more prone to errors. Regional geological sampling surveys may register sample locations by simply plotting them on a topographic map at the time of collection. This assumes that the geologist can accurately identify the field location on the map, and that it is plotted carefully. To enter the locations into a computer, a later digitizing step is also required. This step involves entering reference coordinates and manually positioning a pen or cross-hair at each sample marked on the map. These coordinates are then attached to other data for each sample, using a general-purpose merging program. The sample identifier is entered into both original files, to be used in merging to match the coordinates to the appropriate analytical data.

If the samples are taken at regular intervals along straight lines, it is not necessary to digitize every location. Only the endpoints of the lines, and any points along the line where the direction or sample spacing changes are needed. In this situation, the merging procedure involves linear interpolation between the digitized reference points, based on the sample number (Fig. 2.1)

All of these steps are subject to transcription and placement errors, so it is essential to have the computer replot the map after digitizing and carefully check all samples. Note also that the check plot must include sample identifier as well as position: The coordinates are useless if the wrong sample name is attached.

Scanning systems can also be used for digitizing, similar to their use in entering alphanumeric data. The same type of hardware is used, although possibly scaled up in size to accommodate full-size maps. Software to extract features is also needed, although for maps this problem is considerably more difficult. There may be many conflicting types of information shown on the map, while the digitizing task is to extract only one type. For example, topographic contours may be needed from a map that also shows cultural features, land use, transportation routes and so on. Character recognition also may be needed (e.g., to set the elevation for the contours). It may also be difficult to achieve automatically, since many other bits of text may be close to the desired ones. As a result of these problems, automatic digitizing is not as common as text scanning, and a higher level of operator training is needed to use these systems.

Gridding

Many of the mathematical and graphical operations applied to spatial data assume that the data locations are spaced uniformly in the X and Y directions. The array of locations is called a *grid*. The term *grid* is used to indicate both the array of locations and the associated data values.

The main benefit of regular spacing is computational efficiency. It allows the coordinates to be effectively ignored when doing transformations, filtering, tracing contours, and a variety of other tasks. All of the techniques of matrix manipulation can be applied to spatial data on a

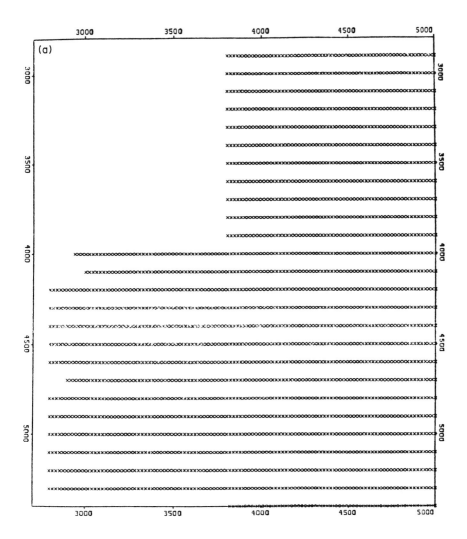

Figure 2.1 Required control points for line data.
A: Nominal regular layout of sampling points for geophysical survey in mountainous terrain. Coordinates were attached to each reading based on this grid.
B: Actual locations for ends of lines and deflection points, determined by later surveying.
C: New coordinates are assigned to all sample locations, by interpolating between control points.

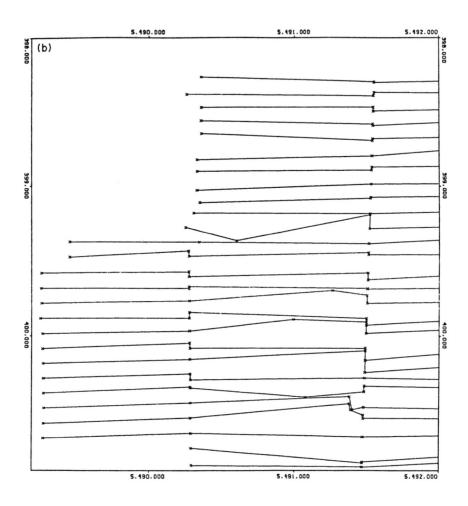

Figure 2.1 Continued.

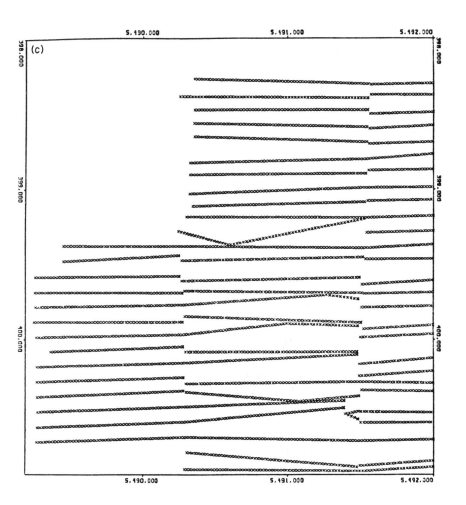

Figure 2.1 Continued.

grid, by storing the values in a two-dimensional array with a set number of rows and columns. Another benefit in working with data on grids is that it becomes easy to make comparisons between different variables, by matching the values for corresponding grid cells. One- and three-dimensional arrays are also used in spatial analysis, for the same reasons.

It is difficult to acquire data on a regular grid, even when the survey is taken along lines. Errors in location data, the need to avoid topographic obstacles, and other factors arise almost inevitably. In many situations, the logistics of doing the survey make it impossible to even attempt sampling on a grid. For example, drilling offshore oil fields often involves several wells fanning in different directions from the same platform. The nature of the survey may also preclude sampling on a grid. Geochemical studies of stream sediments by definition must be restricted to sample sites along existing streams, for example.

To take advantage of the grid format, it is then necessary to transform the original data locations into a regular spacing. This process is called gridding. It may sound like a routine procedure, but in practice it is often a major task. Many different gridding algorithms are available, but none has been found to be universally acceptable, partly because the judgment of whether a gridding program has produced acceptable results is subjective. It depends on the preference of the analyst, the use(s) to which the gridded data will be put, and other factors. Statistical tests can easily indicate whether the basic parameters of the original and gridded data are the same, but local variations between them are often the most important factor.

All gridding methods rely on the assumption of continuity in the data; that is, data values change smoothly between locations. Thus, each grid value can be determined from a relatively small number of the nearest data points. A gridding program normally sets up an array of locations, then estimates the data value for each in turn after searching through the original data for the samples closest to the grid point. In theory, all original data might have some bearing on the estimated grid value; but using all data in gridding is a practical impossibility even with today's fastest computers. In reality, the contributions of distant points are so small that ignoring them makes no real difference. (This effect is a basic principle of geostatistics, which will be discussed briefly in Chapter 7.)

Although the actual calculation of each grid point value may be different, most gridding programs have many common features. First, they must search through the input data to find a set of nearby points to use. This process can be a very time-consuming, so considerable effort in programming is expended to avoid having to do this search by brute force: The program should not have to calculate the distance to every data point for every grid point. Various sorting schemes may be used to organize the input data prior to searching. A maximum distance to search is usually a required parameter, which is used to group the input data. It then is easy to determine whether all samples in a group are beyond the maximum distance. Gridding proceeds through the array of grid-point locations in a systematic fashion, that is, along rows or columns. This procedure allows some additional shortcuts to be taken in selection; since many of the selected points for each grid point may be used for the next one. Visualize the location being gridded as a moving circle: As it advances to the next position, some original data points drop out, while some new ones enter the circle. If directional trends exist in the data, the search circle may be replaced by an ellipse with its major axis aligned with the trend.

Another common aspect is that the stability of the gridding calculation depends on having original data points on all sides. That is, gridding is more reliable in interpolation between points than in extrapolation beyond the bounds of the data. To ensure a good distribution of points in all directions, the area of influence is divided into angular sectors, and the nearest points in each sector are chosen (Fig. 2.2). For example, the search may select the closest four points in each octant of the search circle, instead of the nearest thirty-two points in any direction. This method reduces problems associated with variable data density, which otherwise could have the grid value estimated mainly from a single cluster of data.

After points have been selected, the grid value is calculated. For many methods, this process reduces to computing a linear combination of each point by some weighting function. The weights might be coefficients of a polynomial fit, linear regression coefficients for fitting a plane, some function of distance and so on. Following the assumption of continuity, normally the weighting mechanism places more emphasis on the nearest points. A common method is to use inverse distance weights;

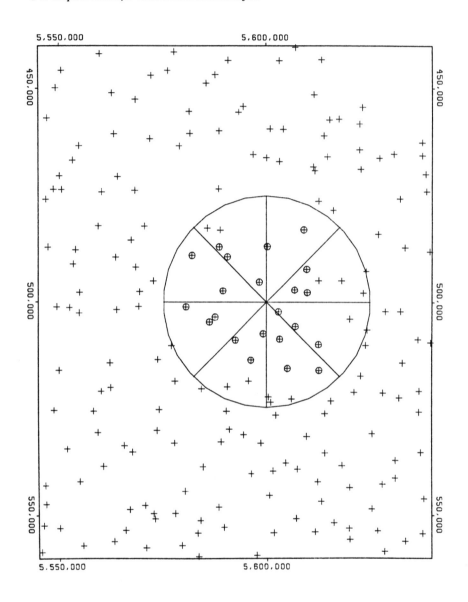

Figure 2.2 Search pattern used in gridding.
Square shows the location of grid cell to be estimated, from data points
marked by +. All data points within circle are tested, and those selected
for gridding are circled.

that is, each data value is divided by the distance from the grid point. (Variations such as inverse distance squared may be used to add greater emphasis to the nearest points.) In the extreme case where a data value falls exactly on the grid location, the grid value can be set to the data value, if the grid is intended to honor the data. This procedure is not always a requirement: For some applications an average value may be more appropriate. The choice of weighting function will determine whether the grid will honor individual data values.

The gridding procedure may use mathematical models of a surface to calculate values at the grid point, for example high-order polynomial fits. Interpolation schemes such as bicubic splines are routinely used. The algorithm may constrain the surface to have some particular properties, such as minimum curvature. Geostatistical estimation methods are also used.

Many different gridding techniques have been developed, although no single method is clearly the best in all circumstances. Gridding success is highly dependent on the nature and distribution of the input data. Several different gridding algorithms should be provided in any contouring package, so that alternatives can be tried easily when the user's preferred method fails. Composite methods may be a good compromise to use as the normal method, which means iteratively reestimating the grid values with different methods, to avoid potential instabilities in a single pass method. For example, the grid values may be first calculated using a simple least-squares planar fit, and then adjusted with a bicubic spline interpolation to allow for smoother transitions between original data points.

The nature of the data to be gridded may be exploited in the gridding method. For example, data taken at close spacing along lines can be difficult to grid, since the spacing between the lines is much greater. Gridding digitized topographic contours may fail, for similar reasons: The density of points along each contour may be high, so that normal search procedures retrieve from only two adjacent contour lines. Gridding programs involving a two-stage process have been developed for these special cases. For digitized contours, this might involve first selecting a single contour level, and interpolating along it to reduce the data density, before a conventional two-dimensional search.

Many gridding programs allow several methods of actually computing the value. Because a large part of the computational effort is in the searching, it is possible to create more than one grid in one pass through the original data. This option may be useful when testing which method does the best job on a particular data set.

Other parameters must also be set by the user. The number of points to retain, maximum distance to search, and criteria on using incomplete data are may be determined by the program, but can be predefined by the user in special circumstances. For example, with large and closely spaced data sets, the automatic choice of maximum searching distance may be larger than required, leading to unnecessarily long computing time. Most programs have limitations on the number of input data points and number of grid cells. These limits may force some preprocessing to reduce the size of the data set, or require creating several grids to cover a large area.

A key parameter is the size of grid cell to use, that is, the spacing between the grid point locations. This spacing may be predetermined by the application: The grid may be needed for input to a modeling program, for which the desired cell size has already been set. Some programs compute the average separation of points to set the grid interval, but this can be tricky when the data distribution is erratic. If the major objective of the grid is for data display, such as contouring, the scale of the display is the most important consideration. The cell size should be small enough to show the desired level of detail in the data, which is often a compromise when program limitations and computing time are considered. The desired cell size (say 10 m) might imply a grid of 500 rows by 800 columns, and the gridding program might be limited to 400 by 400. In addition, note that cutting the grid interval in half results in four times as many cells, which means about four times as long to compute the grid. Later steps using the grid will also run four times longer.

A useful rule of thumb in gridding for a contour map is to use a grid interval equivalent to between half a centimeter and a centimeter at the desired scale, depending on the data spacing. Smaller size (even if the data density warrants it) will not be distinguishable easily. For example, on a 1:10,000 map, the interval would be 50 to 100 meters. For color or shaded maps, more fine detail is distinguishable, so a grid interval equivalent to the data spacing should be used, if all of the detail in the data is to appear on the map. The limitations discussed above still apply, of course.

Although gridding has been discussed as a data preparation step, it is used throughout the analysis and interpretation phases as well. Many derived quantities are created at later stages, to be manipulated and displayed from grid files. Later chapters will discuss some of the specific applications.

BASIC DATA ANALYSIS AND DISPLAY

Although each type of exploration data has its own special characteristics and specialized techniques for data analysis, there are some fundamental tools that can be applied to almost all types. To use a computer effectively in exploration, it is essential to start with these basics. Since they will be used on a variety of data, it is also important to stress ease of use. The basic programs for an analysis system will likely be used much more frequently than the more sophisticated and specialized programs tailored to a single type of data.

Simple Statistics

Statistical analysis is involved almost universally in exploration. At the simplest level, it may involve determining the mean or median value to identify samples with values higher or lower than average. This process in itself can lead to considerable complication, as an average can be calculated in several ways (none of which is inherently more correct than another). Beyond average values, many measures may be used to indicate the range of variation of a group of samples. When the statistics are applied to several variables, or to spatial coordinates and observed data, the situation rapidly becomes more complicated (and beyond the scope of simple statistics that will be widely used).

A program for basic statistics should have the following capabilities:

Accept data in various formats (both in number and type of data fields, and in the physical layout of the data in the computer file)

Compute standard measures for average value and variability:
arithmetic mean, median, geometric mean variance/standard
deviation dispersion

Produce simple graphical displays of various fields: histograms,
two-variable scatter plots

Calculate correlation coefficients between pairs of variables

Apply data transformations, especially logarithmic transforms

Select subsets of data according to data values:
values above or below a threshold
values of two variables within a cluster on scatter plot
match to nonnumeric variables

Apply nonparametric tests

Use a variety of display formats:
normal histograms, exploratory data analysis

Produce graphical displays quickly and interactively

With a program of this type, all types of data can be scanned
routinely as soon as they are entered into the computer. With some
experience with each type, much information can be gained immediately. For example, the mean value may show that a group of samples
is higher than all previous groups in an area. Comparison of standard
deviations can indicate a different degree of variability. We might note
that these differences may be more apparent in the coefficient of
variation, which is the ratio of the standard deviation to the mean.
Going beyond the global parameters of the group, histograms and
scatter plots allow the entire distribution of values to be seen in a
compact form. The need for data selection arises here: It is essential to
easily find the samples that lie in interesting parts of these distributions.
For example, a group of high values may show on a histogram, or a
distinct cluster of points on a scatter plot. The ability to sort out these

values is the first step in checking whether they are errors or legitimate anomalies.

In many geoscience applications, the data to be analyzed have several independent variables. Examining them one at a time can become tedious, so a useful option in a basic statistics program is to be able to create displays for all variables simultaneously. The same approach is needed when examining multiple groups of data selected from a larger set. The program must scan the data to determine some reasonable default parameters to use. Effective output demands a reduced size display, so that multiple histograms, scatter plots, and so forth are shown on the same page (Fig. 2.3). The analyst then can see many variables at the same time, allowing the powers of human vision to quickly find those that are unusual. As a general rule, the more data that can be seen at once, the greater the insights that may be derived from the data.

Computer-Generated Maps

As we have noted already, the ability to draw maps is a key to studying the spatial characteristics of a set of data. In the simplest form, raw data values are posted on a map for visual detection of groups of abnormal samples. If such anomalous points tend to cluster on the map, it is likely that they are related to a common cause. If, however, they appear to be randomly scattered among other samples, it is at least considerably more likely that errors are present. Computer plotting can do better than simply posting numbers on the map, however. Other forms of data display may be used to emphasize anomalous regions.

Contours of a numeric variable are a standard form that predates the computer era, but is still highly effective in many cases. Contouring is based on the assumption that the data values are continuous, that is, there is a smooth transition of values between data locations. Contouring programs require an interpolation of data values onto a rectangular (or triangular) grid system. Gridding can be a difficult task, as noted earlier in this chapter. As a result, computer contouring does not always produce an acceptable representation of the data, as we shall see in later chapters.

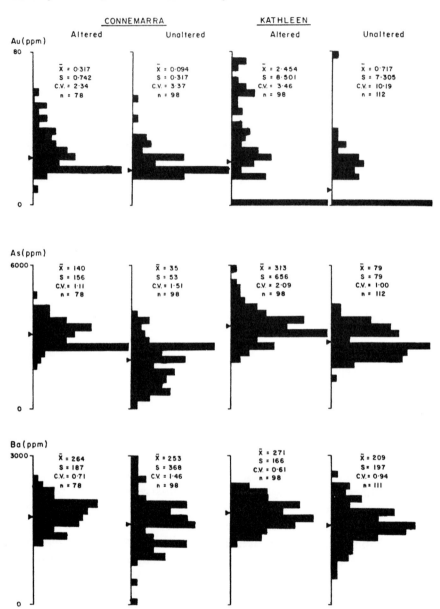

Figure 2.3 A histogram matrix for selected groups of data.
(Reprinted from *Journal of Geochemical Exploration*, vol. 31, p. 244, 1989,
"Discrimination between altered and unaltered rocks at the Connemara
and Kathleen AU deposits, Western Australia," by A.J. Eggo and M.G.
Doepel.)

Triangulation schemes for contouring are becoming common as well. They require creating a set of triangles of different sizes, with each data point at a vertex (or in some cases at the center of the triangle), as shown in Figure 2.4. The end result is a set of cells with known values for the variable at the corners. This method avoids many of the problems of grid interpolation, since original data values are used. It eliminates the possibility of using the interpolated data for array operations, however.

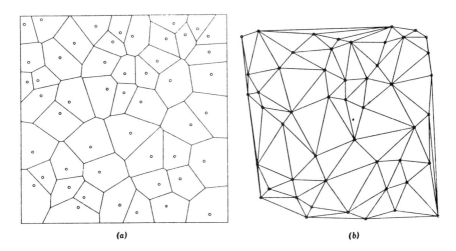

(a)   *(b)*

Figure 2.4 Triangulation network for set of random locations.
A: Tessellation to define areas of influence.
B: Network connecting points with a shared boundary.
(Reprinted from *Spatial Statistics*, by Brian D. Ripley, p. 39, New York, 1981; copyright © 1981 by John Wiley & Sons, Inc.)

For either type, contour tracing is then a matter of linear interpolation along the sides of all cells, to determine the crossing points of the chosen contour levels. These points are then simply connected by lines (for aesthetic reasons, spline fits or other techniques for allowing smooth curvature are applied in the plotting stage). Additional options are available in many contouring programs to label the contour levels, add marks to indicate slope, suppress lines when contours come close together, and perform other embellishments to make the computer output comparable to a manually drawn map.

Different symbols may be used to represent different classes of data. This technique also can be applied to nonnumeric data. Varying the size as well as the form of the symbols can make visual discrimination easier. With recent improvements in technology, computers can readily produce color displays, which are even easier to interpret. All of these forms can be combined with posted numbers, so that the basic data can be compared to the symbolic representation. (Examples of many different computer plots will be seen in later chapters.)

To make routine use of such displays, a flexible general-purpose mapping program is required. Like the statistics program, a primary requirement is that it accept a variety of data and formats. It also must have many options for plotting data (as noted previously), and allow easy transition from one form to another. Since it may take a considerable amount of time to produce a plot (both in straight computation and in actually drawing the map on a plotter), it is also desirable to have a convenient way to define several maps at once, without having to wait for each one to be completed.

The data selection procedure described in conjunction with basic statistics is useful when examining a map. If a contoured display outlines some anomalous areas, the samples in the area can be retrieved by a threshold test. Symbolic maps may show local clusters of similar samples, which might be selected by drawing a polygon around the cluster, and digitizing its outline for use by the selection program. Selection might be done interactively, by drawing the polygon directly over a display on the screen.

Other Types of Graphical Display

Although maps of various forms are the usual graphical format in exploration data analysis, some other forms are also widely used for specific types of data. Examples of many of these will be given in later chapters.

Most cases of departure from the map format are for studying data that do not have two spatial coordinates. When there is a single coordinate, the data variable (or variables) generally are plotted in a profile form, with the horizontal axis representing the coordinate, and the vertical axis the data value. Individual samples are connected by straight lines, to emphasize the variations at adjacent points. If there are several variables to examine concurrently, they may plotted on the same graph, using different line styles or colors to distinguish each one. Alternately, they may be shown as separate profiles with the same horizontal scale and a limited vertical extent to allow the individual plots to be stacked above each other for visual comparison.

There is more to the analysis of one-dimensional data than graphical display. Since the independent variable is often time, many special analytical techniques have been developed under the general classification of time-series analysis. For example, consider observations of temperatures over a year or seismic wave amplitude recorded for several seconds after inducing a wave. Many of the operations applied to time series are plotted in the same form as the original data, so that a primary requirement of a display program for one-dimensional data is that it can apply operations such as correlations, derivatives, polynomial fits, and others.

Time-series methods (analytical and graphical) are not restricted to cases of data measured at fixed increments of time. The independent variable may be any unit of measurement. In exploration, the most likely alternate to time is linear distance along a sampling traverse (which could be a road, a cut line through a forest, a drillhole into the Earth, or an aircraft's flight path, among others). Wherever possible, samples are taken at a constant increment of time or distance, to simplify the computations needed in analysis. This feature is not essential in time-series analysis, however (especially if the analytical method is largely graphical). Sampling is often irregular, particularly if there are natural constraints on the sampling procedure. For example, a time series of magnitudes of large earthquakes can only specify the observed times (there is no control on the time of each event). In logging a drillhole, samples may be taken only from specific types of rock, which occur at erratic positions in the hole. If the analytical procedure requires a regular interval, the original data values are interpolated. One-dimensional methods such as spline fits fortunately are less difficult to implement and more stable than the gridding methods needed for two-dimensional data.

The other variation from the two-dimensional case is when there are three or more coordinates. As we have noted already, when working with volumes of the Earth there are three spatial coordinates, but the two-dimensional forms of maps are generally used, by plotting a series of slices through the volume containing all the data. This is the result of the difficulty in showing all three coordinates on a two-dimensional plot.

In studying data with several variables, special graphical forms may be used in which the coordinates of the plot are various combinations of the variables. Two variables are shown in a scatter plot, by simply plotting a symbol at each point defined by a pair of values for the chosen variables. The symbol may be used to indicate the value of a third variable, or may represent the number of samples at that point (when the scale of the plot does not allow each sample to be distinguished easily).

A ternary diagram is the three-dimensional equivalent of the scatter plot. Points are plotted within a triangle by scaling distance from each corner according to the values of three variables. Similar to the scatter plot, the symbol plotted at each position may carry other information. Also like the scatter plot (and indeed all other methods for displaying multivariate data), the primary objective of creating the plot is to identify trends and common relationships among the samples.

To show more than three variables on a plot is difficult. Most techniques for multidimensional analysis involve plotting a special figure for each sample, with the lengths of various parts of the figure scaled by the individual data values. For example, six variables might be shown as the arms of a six-sided star. Other forms applied in various types of data analysis are Chernoff faces, trees, rose diagrams, and bar diagrams. These methods obviously will succeed for only a small number of samples, as the method relies on visual examination to group samples by similarity of the figures. Computer algorithms for pattern recognition may be employed as a first cut at segregation into groups, but the uncertainty in defining significant patterns demands manual verification in almost all situations.

Any of the special multivariate plots can be combined with the map format, by plotting each special figure at its location on the map. This procedure can be useful to see if visual groupings also show a spatial dependence (i.e., do figures of common shape tend to appear in adjacent locations?). A map-drawing program requires considerable enhancements to plot special figures, so the process of plotting the figures on a map is frequently done manually. Given the practical consideration of only applying these techniques to small sets of data, this is generally not a major chore.

USING COMPUTERS IN REMOTE LOCATIONS

One of the major developments in exploration in the 1980s is the widespread use of portable computers. Many types of data now are directly recorded in computer format, or typed into a computer file immediately. The availability of portable, rugged, and inexpensive computers was the driving force. With the ever-increasing power of these machines, more sophisticated types of data processing and analysis now are done at the exploration site.

Taking a computer to the field is not without problems, particularly if the intent is to provide the same capabilities as a larger computer at the home office. Portable computers tend to have less memory and disk storage space than office machines, so it may be necessary to restrict the size of programs and data files. This limitation is not necessarily a major problem, since other factors may limit the type of computing that can be done in the field.

In geochemical surveys, for example, laboratory results may not be available for several weeks after sample collection, by which time the geologist may be back in the office. In addition, the demands of getting the field work done leave little time for detailed data analysis, which is left as a separate task to be done later. Another factor is that the data-analysis phase may require additional equipment (such as a large plotter), which cannot be set up easily in a field camp.

Another difficulty in using computers effectively in the field is communications. The field computer may be used to receive data from other computers, or to send data to other locations. Most analytical laboratories have an option to send results in computer form over a telephone line. Where the field computer is collecting data for analysis on a different machine in the office, there may be considerable time savings if the data can be immediately transmitted back to the office. Computer connections are also useful for relaying messages between the office and the field. However, in remote locations, telephone lines may not be available, or may be so noisy that computer communication is impractical. Radio or satellite communications can be used, although the expense of setting up the equipment may exclude their use on small, short-term projects.

In principle, any computer should be able to communicate with any other. In practice, however, this is often quite difficult, since there are no global standards for communications protocols or error-checking procedures. To have one computer talk to another then requires specific details on the coding scheme of each machine: for example, the number of bits used to encode a character; the number of start, stop, and parity bits transmitted with each character; the speed of transmission; and other parameters. If either computer makes the wrong assumptions about the values of these parameters, a usable connection will be impossible.

The ability to adapt communications parameters when connecting different computers is a function of software. Many packages are available to provide the basics for almost any type of computer to act as a terminal when connected to another. This means that any typed characters are simply transmitted to the communications port (and on to the other computer), and all incoming transmissions from the other end are simply displayed on the screen. Control of the remote computer is taken over by the machine at the other end of the communications line.

Simple terminal mode is fine if the link between computers is used only to receive or send short messages. More sophisticated software is needed to transfer data files, however. In this case, both computers must be processing incoming and outgoing data at the same time. Data to be transmitted must be read from a file, and encoded with some form of error-detection value before being sent. The error code takes various forms, but always is dependent only on the data being transmitted. The receiving computer thus can recalculate the code and check the value against that received with the data. If the codes do not match, a signal to try again is sent.

Because there is no standard way of doing the error check, it is essential to have compatible software on all computers that are expected to transfer files between each other. Variations in how to signal acceptance or rejection of data, the size of blocks of data in each transmission, and other possible differences between systems add to the problem. In general, if all machines use the same operating system (e.g., MS-DOS for personal computers), it should be easy to install a common communications package. When different operating systems are involved, especially involving traditional mainframe computers, it is more difficult to locate compatible software to run on all machines. This is no longer a major problem, however, as a number of effective data transmission protocols now are commonly available on many different types of computers. Most personal computer communications programs include several different transfer methods, which can be selected from a menu to match the software running on the other computer.

CHAPTER 3

Geochemical
Data Analysis

The discussion of specific exploration methods begins with geo-chemical surveys. Although many of the analytical techniques are used in other fields, the main reason for looking first at geochemistry is that it usually is confined to sampling the surface only. Therefore the spatial analysis is strictly two-dimensional, and we can defer introducing the complications of three dimensions until later.

Throughout this chapter, the word *element* will be used to describe the variables subjected to geochemical analysis. While this term often means the chemical elements, all of the discussion applies equally well to compounds such as hydrocarbons or complex pollutants. For simplicity, the term will be assumed to include the broader context.

THE NATURE OF GEOCHEMICAL DATA

Most of the analytical and graphical methods applied in exploration make a tacit assumption that the data are reproducible: Taking the same measurement at the same location results in the same value. Geochemical surveys have some unique problems that make this assumption questionable, at least within the practical limits of what can be achieved in the field. Perhaps more than with any other type of data, the uncertainties inherent in the sampling and measurement procedures must be acknowledged in making an interpretation.

The first thing to keep in mind about geochemical data is that the values generally represent minute fractions of the total sample. Depending on the element, typical background levels may range from a few parts per billion to a few parts per million. At the lower end of the range, it is possible that a few microscopic grains of a mineral can create a significant anomaly. Thus the volume of material in each sample must be taken into account in judging the validity of any abnormal values. The mode of occurrence of the particular element is also important, especially for those such as gold and silver, which can occur both in erratic particles of the native element and bound in various minerals.

Although the actual sample is small, it is assumed to represent a large (sometimes vast) area of the surface. A fundamental principle behind the success of geochemical exploration is that trace quantities of elements in a mineralized zone are spread throughout the surrounding region by erosion, leaching, and so on. In regional surveys, this principle implies that the source of a geochemical anomaly could lie anywhere in the area drained by the stream where the sample was taken. In following up on regional surveys, detailed grids of soil samples are used to define potential source areas. Planning the grid requires an implicit assumption that the geochemical halo of the source will be at least as large as the spacing between samples. An understanding of the transport mechanisms for different elements is needed to estimate what the size of a halo might be.

The same principle applies in environmental testing of streams and lakes. Here the problem is to determine how far a pollutant might travel before it is diluted sufficiently to be undetectable. In addition, the source may be intermittent, so the surveying might have to be repeated at different times. Another complicating factor is that particular types of samples are available only in certain areas; for example, sampling lake sediments obviously restricts the locations much more than sampling soils.

With erratic distributions in the Earth, it is difficult to ensure that a major anomaly has not been missed even by detailed sampling. The chances of detecting a source of mineralization are dependent on the particular elements involved, since different elements may have widely different distribution patterns, even when coming from the same source. The mobility of elements depends on how easily they can be dissolved, their chemical affinity to other elements, the environment of the area, and other factors. Note also that a surface soil sample will not necessarily detect the same source as a stream sediment, because the trace elements may enter the stream by underground drainage, while there is little vertical migration to the surface. The end result is that a single anomalous sample may be highly significant, even if surrounded by samples that contain only background levels. Thus, it is difficult to detect errors by spatial analysis, since the level of continuity may be lower than with most other types of data.

To get around this problem, geochemical surveys usually involve analyzing samples for several elements, some of which may serve as pathfinders for the eventual target elements. To make effective use of such multielement data requires both a knowledge of their chemical characteristics, and a model of mineral associations in the particular geologic setting. It also requires an understanding of multivariate statistics, and software to perform the necessary computations.

In addition to these fundamental problems, there are limitations in the analytical methods used to determine the concentrations of the elements. For any element, there may be several different measurement techniques, all of which may be suitable for some (but not all) types of samples. The different analytical methods are subject to detection limits (i.e., they cannot measure concentrations below some minimum level), and to constraints on the sampling and sample preparation procedures. Because of costs and other operational considerations, the optimum technique for a particular element may not be used, in favor of another that is more widely applicable. The method of analysis is chosen to give acceptable results for several elements at the same time, rather than running an independent analysis for each element. Another consideration in interpreting the results of analysis is that many techniques measure only partial concentrations of each element.

Most geochemical measurements have a high level of uncertainty in the values. Especially when concentrations are low (i.e., near the background level, which also may be near the detection limit), the measured values easily may have errors of 50 percent or more. This makes the detection of subtle anomalies difficult, considering the possibility that the anomaly will appear in only a few samples.

VALIDATING GEOCHEMICAL DATA

Because of the high uncertainty associated with geochemical analyses, more effort has to be put into verifying that the data are correct than for most other types of exploration. Since much of the uncertainty is inherent in the method, it is never possible to attain 100 percent confidence in the data.

The primary method for improving the confidence level is to acquire duplicate data. This may be achieved in two ways: by collecting duplicate samples in the field, and by running separate analyses on separate portions of one sample. These techniques are not equivalent, as they measure two different aspects of the variability of the data.

Taking extra samples at the same location in the field gives a measure of the erratic distribution of trace elements on local scales. The measured values can also be influenced by the sampling method itself, and by the experience and technique of the person taking the sample. To check these potential problems, it is obvious that several samples must be taken, using more than one method and more than one sampler. It is not practical to collect all the extra samples at every location, but it is essential to take enough of them to give reliable statistics. The duplicates also should provide a reasonable coverage of the entire survey region, to avoid bias by doing all the checks in one area, which may itself be anomalous.

The other major source of uncertainty is the laboratory analysis of the sample. Duplication of lab assays on the same sample is the most direct measure of repeatability of the results. Here too there are several variations on the theme. The original sample may be split before sample preparation, which tests the entire laboratory procedure (and to some extent geologic variability, since it involves separate source material). To test the inherent variation in the analytical instrument, the final sample material may simply be run through the instrument twice. The sample preparation stage reduces the volume of material to a minute amount (by crushing and grinding the sample, and then dissolving it in an acid, for example), so all of the prepared material should be equivalent.

Other complications in the laboratory are different analytical methods (atomic absorption spectroscopy, neutron activation, X-ray fluorescence, to name a few). From experience with the elements of interest and the type of sample material, a preferred method is usually known at the outset of the survey. Running selected samples through another procedure may help to point out biases, and can help to determine whether a high variability of results is real or at least partly due to the analytical method. If two independent methods show the same degree of variation, it is more likely that the cause is geologic (although this is not certain).

A common technique for testing the laboratory procedure as a whole is to send sample splits to different laboratories. The closer the agreement between the results, the greater the confidence in the geologic interpretation based on the data. For reasons of cost, this method usually is applied to a subset of the samples, although once again they must be selected to be representative of the complete set. In some situations, all samples may be analyzed twice, for example, if a project is in a late stage of development and a major financial decision depends on the results.

Regardless of how the duplicate results were obtained, or what they represent, the methods for studying them are the same. The starting point is to compute the basic statistics of each set of data (mean, standard deviation, etc., for each variable of interest). Visual inspection then can quickly point out gross differences, although this can be misleading. To correctly judge differences that are statistically significant is not always a simple matter, because the number of samples in each set of data is a critical factor. (Unless one always works with data sets of about the same size, experience is not always helpful.)

The most widely used statistical methods for comparing duplicate samples require the assumption that the variables have a normal distribution. For most geochemical analyses, this is in the form of the lognornmal distribution (i.e., the logarithms of the data values have a normal distribution), due to the skewed nature of element concentrations. The majority of samples will have low values, while a few have much higher values. Since the value must be positive, the distribution of arithmetic values cannot be symmetrical, which would require corresponding large negative values.

Although there are theoretical arguments against using the assumption of normality, it has been well established empirically that many numerical parameters of the Earth have this characteristic. Given the past history of success on many different mathematical problems, the normal distribution will likely hold its place as the guiding principle of many statistics applications.

In the past, the complex calculations involved in other forms of statistical analysis discouraged widespread use. Nonparametric methods which do not require an assumption about *any* type of distribution have been limited. A lack of experience with the methods is also a contributing factor. With present-day computers and readily available software, it is likely that alternate methods will be used more widely. If there is any doubt about the assumption of normality, nonparametric analysis should be tested (see the Bibliography for sources of more information).

Treating the duplicate samples as two different measures of the same unknown, the first question is whether any observed differences are statistically significant. Assuming normal distributions, or the application of transformations such as logarithms to produce a normally distributed variable, the answer is obtained using the T-test. The T statistic can be calculated simply knowing the mean and standard distribution, and compared to tables of the student T distribution. This distribution is the equivalent of the normal distribution, allowing for the number of samples available to calculate the mean and standard deviation. In addition to depending on the number of samples, the T distribution also is related to a significance level, which might be viewed as the probability of error. (Unless there are an infinite number of samples, all computed values such as mean and standard deviation are simply estimates of an unknown, always with some chance of being in error.) The result of the T-test is a statement that the means of each group are (or are not) different at the stated level of significance (typically 95%).

The application of the T-test to the means of the two sets of data simply indicates whether the mean values are significantly different. This constraint is loose, and for small data sets does not provide much information. This calculation does not exploit the fact that the samples are duplicates, however. It is just as valid when applied to two independent sets of data. For example, two subsets of a large group of samples might be selected to see if mean values of the elemental concentrations are related to some other variable.

The paired T-test is used on the duplicate samples, which in effect tests whether the average difference between sample values is significantly different from zero. Like the simple T-test, the results are expressed at some significance level. As the name implies, data values are treated as pairs, and thus must be stored in a form where they can be matched easily to compute the differences between values. The simplest form is to have both values as fields on the same record of a data file.

A widely used nonparametric test for duplicate samples is the paired-rank test (also called the Wilcoxon test). It does not use data values or differences directly, but instead sorts both sets of values into their rank order. The differences in sequencing between the sets then are used to measure the similarity of the series. The basic principle is that the groups are not significantly different if the rankings are similar (i.e., samples that have low values in one group also have low values in the other). Like the T-test, the power of the test depends on the number of samples. The more duplicate samples that are available, the more certain we can be that the duplicates are statistically the same (or different).

A more descriptive way of examining duplicates is to plot both values on a standard X-Y scatter plot. It is useful to have the same limits on each axis, and to show the regression line for the data in comparison to the 1:1 line (i.e., the expected regression line if there is no bias in either set of data). The scatter plot also may illuminate details of the relationship that cannot be indicated by a simple measure such as a T-test. It may be that more than one bias appears, one for high values and one for low, for example. The techniques of exploratory data analysis also include a variety of compact graphical forms for comparing sets of data. Some examples of scatter plots are shown in Figure 3.1.

When the geochemical analyses involve several elements, the entire process of analyzing duplicate samples is repeated for each element. Repetition is necessary because of different physical character-

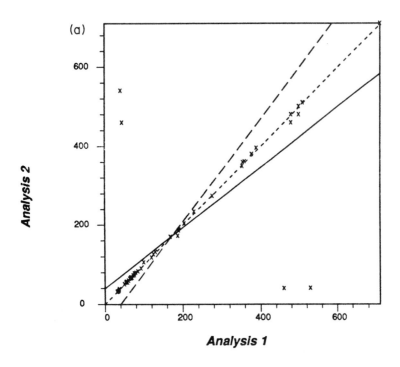

(a)

Analysis 1

Number of data   61

Correlation coefficient = 0.768   T Statistic (for different means) =   0.0492

LS y on x: Y =   39.5611 +   0.7658 X   Mean and Variance of X: 172.09   29710.1
LS x on y: X =   40.0841 +   0.7704 Y   Mean and Variance of Y: 171.36   29534.6
UNBIASED: Y =   -0.0753 +   0.9962 X

Figure 3.1 Scatter plots for paired data.
   A: Duplicate laboratory analyses for gold in rock samples. Note the
supporting statistical information, such as the correlation coefficient and
regression line fit. Most points lie on the 1:1 line, but some outliers cause
a bias in the regression line.
   B: Analyses of trace elements in different types of plants. Note use of
symbols to show the soil type.
   (*B* reprinted from *Applied Environmental Geochemistry*, ed. I. Thornton,
Academic Press, 1983, p. 108, "Soils and plants in the geochemical
environment, " by J. Kubota.)

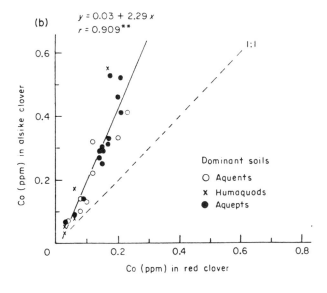

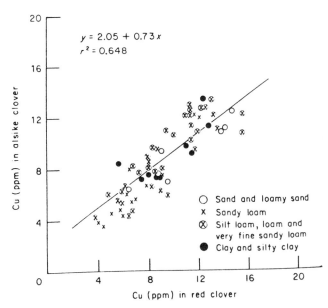

Figure 3.1 Continued.

istics of each element, and because of different laboratory procedures that may be involved. The end result may be that some elements will be given higher confidence ratings than others in later study of the complete set of data.

SPATIAL ANALYSIS AND MAPPING

Most geochemical surveys are studied by graphical methods, rather than analytic methods such as geostatistics. This is partly due to the inherent uncertainties in geochemical data, which preclude drawing firm statistical conclusions. In addition, the extra expense of detailed geostatistical analysis cannot be justified at the early stages of exploration: The data are generally too sparse to get reliable results. It is also partly due to a lack of experience in applying geostatistics to geochemical problems, a factor that is likely to change in the future.

The main objective of many geochemical surveys is to define a target area for more detailed exploration (either by additional geochemical sampling or by other methods). The key to defining an area graphically is to use a form of displaying data that emphasizes anomalous areas. In multielement surveys this problem can become complex, since it is desirable to look for regions that are anomalous in more than one element. Combining displays of several variables on one map is necessary, which can easily lead to uninterpretable clutter, unless the form of display is carefully planned.

Contouring the values of a single variable is a common method of outlining anomalies. Use of this technique stems as much from its simplicity as from its inherent ability to give the desired result. Since contouring is widely applied to many types of data, and is part of almost all computer mapping systems, it is often the first method that is tried. There is a potential problem with the contour format, in that a closed contour gives the impression that the region within it is uniform. On a geochemical map, there may be many closed contours caused by only one or two samples, yet enclosing a large area. This is especially true in regional surveys, where the distribution of samples is erratic.

Another potential problem with contour maps is that contouring algorithms make an assumption of data continuity, which is not always true with geochemical samples. A single sample may have high values for several elements, while all surrounding samples are at the background level. At best, a series of closed contours will form a bull's-eye around the anomalous sample. If the anomalous value is extremely high, it may cause severe problems in the interpolation scheme used to trace contours between the samples. In this situation, the bull's-eye effect may extend far beyond the area around the sample, which can give the misleading impression that a large area is involved, not just a single location. Shaded color maps are sometimes used for geochemical data (as well as other types). They are in effect a form of contouring, so the same limitations and problems apply.

This problem is likely to be more severe with contouring based on a regular grid (the most common method) than with methods employing triangulation between original samples. Gridding generally requires estimating many more intermediate points, and involves more complex mathematical formulae that are more likely to "blow up" when extreme data values are encountered, as we noted in Chapter 2. For samples taken along lines, the chance of gridding problems increases, since the nonuniform distribution of samples and the rapid variations in geochemical values both violate basic assumptions of many gridding algorithms.

Because of the problems with contouring, many geochemists prefer other display methods. The most popular alternative is to use different symbols plotted at each sample location to indicate the data value for that sample. This procedure can involve changes in the shape of the symbol, its size, its degree of blackness, or combinations of all of these parameters. When color output is conveniently available, the color of the symbol may also be used.

Geochemical surveys are aimed at finding anomalous samples, which almost always means those with the highest values. As a result, simple coding schemes that assign different symbols to samples with values in specific ranges are geared to make the highest classes most visually distinctive. This enhancement may be achieved by increasing the size of the symbol according to data value. Even more emphasis is obtained by also changing the shape, or progressively shading in the symbol.

In Figure 3.2, a variety of formats for highlighting anomalous samples are shown. When contours are used, additional emphasis comes from plotting the higher levels with heavier lines (or different colors). With symbolic maps, this option is not usually required, since the variations in the symbol already provide a high level of visual distinction.

Varying more than one symbol parameter is also used to indicate values of another variable. In this case, there are two coding schemes, one perhaps on size for the first variable, and the second employing the shape of the symbol. Using color or line weight allows the possibility of simultaneously plotting a third variable, but in practice it is seldom done, since it is difficult to make such a complex map easily readable.

MULTIVARIATE ANALYSIS

One of the prime characteristics of geochemical surveys is the existence of multiple variables in the analytical data. With modern laboratory instruments, concentrations of twenty or more elements can easily be measured simultaneously. In conjunction with the ready availability of multielement data, improved understanding of the genesis of ore deposits and hydrocarbon reservoirs demands examination of subtle variations among many trace constituents. As a result, there is an increasing need for efficient ways to study interrelationships among many variables.

Mathematical approaches to multivariate analysis are a well-developed branch of statistics. The main problem in application of these techniques is probably that there are too many of them, rather than a lack of acceptable tools. There does not seem to be a general consensus on which methods are the most likely to be useful. The decision is a matter of preference of the individual analyst, including not using statistical methods at all but relying on graphical displays, as we shall see later.

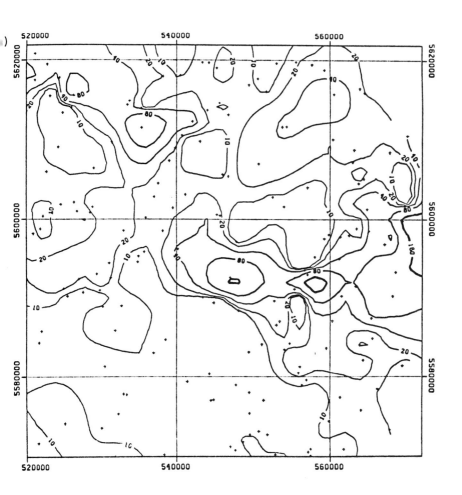

Figure 3.2 Map formats for highlighting anomalous samples.
A: Contours, with higher values marked with heavier lines.
B: Different symbols for several ranges of numerical values.
C: Variable-size symbols for ranges of numerical values.
D: Different line thickness and text size for annotation.

(b)

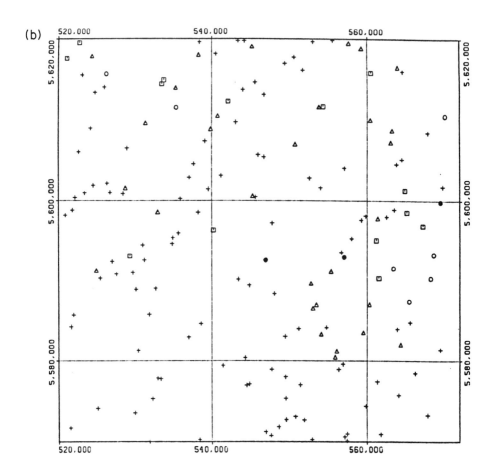

Figure 3.2 Continued.

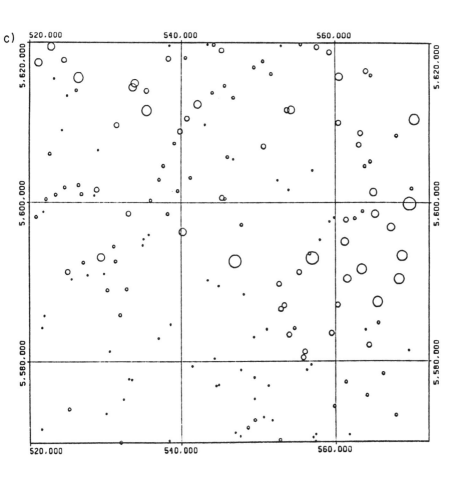

Figure 3.2 Continued.

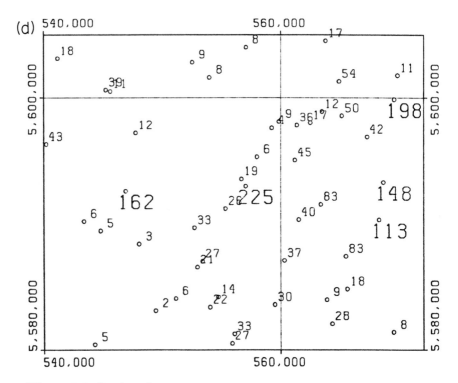

Figure 3.2 Continued.

Graphical Methods

The simplest multivariate situation is having two variables. This case is easily handled by plotting all the samples on an *X-Y* scatter plot. Correlations between the variables will be immediately apparent, as will the existence of any clusters of anomalous samples. Since trace-element concentrations are often skewed, data transformations may be needed to produce an acceptable plot. Note that when using transforms in a scatter plot (or indeed any type of analysis), it may be necessary to transform only one of the variables. This option needs to be considered in any programs designed for general data analysis. The decision to transform only one variable may be based on the observed distributions (as seen in a histogram), or on the nature of each variable (e.g., in water sampling, a trace element may be compared to the pH of the water; pH is by definition a logarithmic measure already).

Logarithmic transforms are the most widely used, for several reasons. First, as noted earlier, geochemical distributions tend to be logarithmic. Thus, all the statistical techniques based on the normal distribution can be applied to the transformed data. Because of the long history of log transforms, most analysts are more familiar with this technique, and will tend to use it as a first choice.

The basic scatter plot is usually supplemented by computing a regression line fit, and from it a correlation coefficient. In most situations a least-squares straight-line fit of Y to X is used. It may yield a deceptive measurement, since the procedure depends on the assumption that Y is a dependent variable to X; that is, X is perfectly known, and all sources of error or variation are in Y. This is obviously not always (or even usually) true. One of many alternatives to the least-squares fit line is the reduced major axis (RMA) line, which allows for variation (or error) in both X and Y.

When there are a number of variables, the first step is to compute a correlation matrix, which consists of the correlation coefficients for all possible pairs of variables. Since the correlation coefficient boils down the entire relationship between each pair to a single number, it often misrepresents the significant features of the relationship. With modern computer capabilities, it is not too difficult to replace the correlation matrix with a scatter-plot matrix (Fig. 3.3). By generating all possible scatter plots, a quick visual scan should allow any anomalous patterns to be detected. To make the output most useful, all plots should be shown on the same page, which entails using computer graphics to display the plots rather than simpler printer displays.

When nonnumeric variables are involved, the scatter-plot form cannot be used. Groups of samples corresponding to each value of the variable (rock type, for example), are selected from the database, and histograms are plotted for each group. The compact matrix format is useful here as well.

As the number of variables increases, the ability to create simple graphical forms or statistical measures rapidly decreases. This is due in part to the inherent complexity of the situation, and also to the fundamental problem of confining displays to a two-dimensional surface. It also stems to some degree from the less frequent need for multivariate analysis, which leads to less emphasis on developing analytical tech-

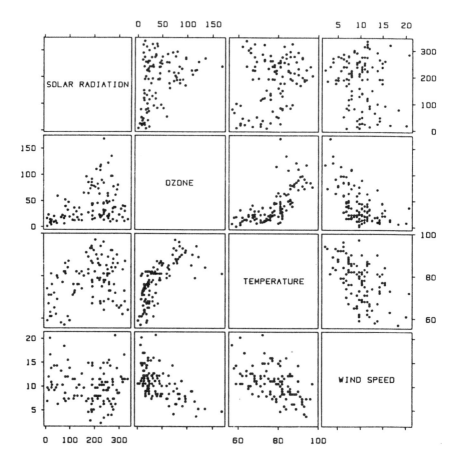

Figure 3.3 Scatter-plot matrix.
(Reprinted from *The Elements of Graphing Data*, by W.S. Cleveland, p.
211. Copyright © 1985 by Bell Telephone Laboratories, Inc., Murray Hill,
New York. Reprinted by permission of Wadsworth and Brooks/Cole
Advanced Book Program, Pacific Grove, CA 93950.)

niques. The result is that it is difficult to gain experience and confidence
in handling multidimensional problems, and simultaneous analysis of
more than two variables is often avoided, even when the situation calls
for it.

When studying three variables, a ternary diagram is a relatively
simple format for graphical display (Fig. 3.4). A program for plotting

these diagrams must provide scaling options to allow for different units of measurement (or ranges of values) for each variable. Without some form of normalizing, all values may plot in one corner of the triangle, and thus fail to provide any visual discrimination of distinct groups of samples.

For more than three variables, direct plotting techniques rely on drawing a characteristic figure for each sample, for example, a multi-armed star. The dimensions of parts of the figure (e.g., the length of each arm of the star) are determined by the values of each variable in the particular sample being plotted. The figures for each sample are then visually separated into groups to identify samples that may be related, or that are anomalous. They might also be plotted on maps, to check for spatial correlations or migration patterns. For example, hydrocarbons from several reservoirs might be traced to a single source.

Some examples of multidimensional figures are shown in Figure 3.5.

While scatter plots and ternary diagrams can readily display large numbers of samples, graphical formats for more than three variables may not be effective for large sets of data. This restriction stems from the use of distinctive shapes controlled by the variables, so that the complete shape must be visible for comparison to other samples. In the scatter-plot format, each sample may be represented by a dot or cross, and a high degree of overplotting is acceptable: The important thing is to be able to see clusters in subareas of the diagram.

Statistical Methods

To study large data sets with many variables, it is necessary to reduce the problem to a smaller number of dimensions. The traditional way to do this is with statistical methods such as principal components or factor analysis. These techniques rely on the assumption that some of the observed variables are interdependent, so that it is possible to compute new variables as linear combinations of the original variables. The result should be fewer variables to examine, allowing use of methods such as scatter plots to search for anomalous groups.

Multivariate statistical methods consider each sample to represent a point in an N-dimensional space, where N is the number of data fields

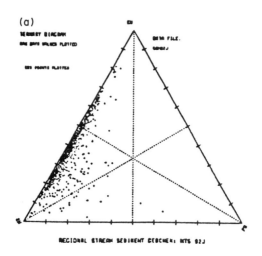

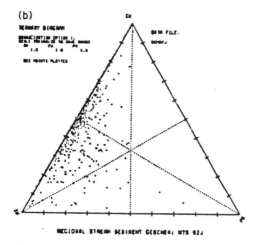

Figure 3.4 Ternary diagram.
  A: No scaling.
  B: Each variable scaled to have a range of 0 to 1.
  C: Each variable scaled to have the same mean value.
  D: Log transformation of each variable.

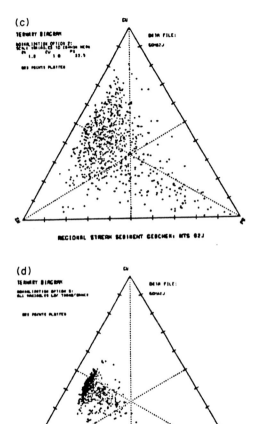

Figure 3.4 Continued.

(a)

Figure 3.5 Multidimensional figures.
A: Star diagrams, trees, and profile symbol plots. Examples are from another field, but the form can easily be applied to exploration data. (Reprinted from *Graphical Methods for Data Analysis*, by J. M. Chambers, W. S. Cleveland, B. Kleiner, and P. A. Tukey, p. 161, 163, and 166. Copyright ©1983 by Bell Telephone Laboratories, Inc., Murray Hill, New York. Reprinted by permission of Wadsworth and Brooks/Cole Advanced Book Program, Pacific Grove, CA 93950.)

(b)

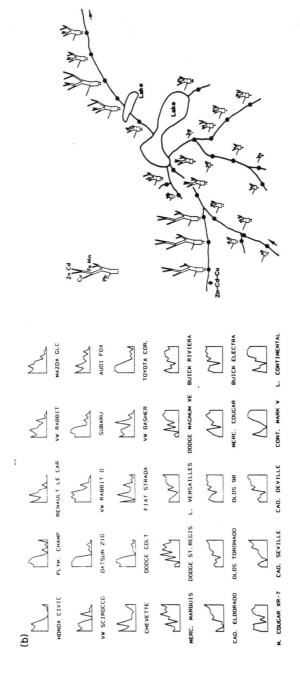

HONDA CIVIC   PLYM. CHAMP   RENAULT LE CAR   VW RABBIT   MAZDA GLC

VW SCIROCCO   DATSUN 210   VW RABBIT D.   SUBARU   AUDI FOX

CHEVETTE   DODGE COLT   FIAT STRADA   VW DASHER   TOYOTA COR.

MERC. MARQUIS   DODGE ST. REGIS   L. VERSAILLES   DODGE MAGNUM XE   BUICK RIVIERA

CAD. ELDORADO   OLDS TORONADO   OLDS 98   MERC. COUGAR   BUICK ELECTRA

M. COUGAR XR-7   CAD. SEVILLE   CAD. DEVILLE   CONT. MARK V   L. CONTINENTAL

B: Trees shown in map orientation. (Reprinted from *Journal of Geochemical Exploration*, vol. 32, p. 337, 1989, "The chi-square plots: a tool for multivariate outlier recognition," by R. G. Garrett.)

associated with each sample. These variables may not represent the most efficient coordinate system for viewing the data. It might be possible to use a smaller number of coordinates by a transformation of the coordinate system (i.e., the original variables). For example, it takes three coordinates to describe position of points within a sphere. If, however, all points lie on a plane that slices through the sphere, transformation of coordinates to a system aligned with the plane allows all points to be represented by two coordinates (since the third one becomes a constant). Even if the points are not planar, a two-dimensional view can be created by projecting each point onto the plane. The objective of multivariate transformations is to find such special orientations of the coordinate system.

The computer methods for solving these problems rely on matrix operations. The starting point is the correlation matrix of all variables, or some near analogue of it. Finding the principal components is equivalent to finding the eigenvalues of the matrix. The reduction in dimensions comes from ranking the eigenvalues, and retaining only those needed to give a desired level of fit to the data. For example, it may turn out that the five largest eigenvalues of a twenty-variable correlation matrix represent 99 percent of the variability in the data, so there would be little need to consider more than five principal components.

Since the prime objective of geochemical exploration is to locate anomalous areas, the next step in multivariate analysis is to determine which samples seem abnormal when considered in terms of the transformed multidimensional coordinates. For example, the values of the principal components would be computed for each sample, and scatter plots would be constructed using pairs of the principal components as axes. This process will require fewer plots than examining all pairs of the original variables. The use of the scatter plot is the same: Look for clusters of points, and select them for further study. The principle in factor analysis is similar, using factor scores for each sample and each computed factor. Note that in use of Q-mode factor analysis, the factors themselves define the related groups of samples, since the analysis is based on the similarity matrix between all samples, rather than the correlations between variables. An alternate method to find anomalies is to examine the higher principal components or factors directly.

Samples with larger values for the higher components are likely to have additional sources of variation than just the regional trends.

Software for multivariate analysis must allow for the many possibilities outlined above. Since there are many common operations to be performed, a package program is a good way to reduce the programming effort, and to simplify using the different methods. The program uses identical procedures for loading the data and storing the results, and allows a simple menu choice of the particular type(s) of analysis. The output results must be in a form suitable for other components of the complete data-analysis system. Computed values such as factor scores are added as new fields to the data base. These new fields can be used for statistical comparisons with other variables, plotting on maps, scanning for anomalies, and so on, that is, all of the procedures used on the other variables (and indeed other types of data).

In some situations, the computed principal components or factors may be used directly in interpretation. Each factor may involve distinct groups of elements, which should be related because of the geochemical mechanisms that produced the mineralization in the first place. Unfortunately in practice the relationships often are not obvious, and the same elements may contribute strongly to more than one component.

Because of the difficulties in interpreting results of the multivariate statistical techniques, and to avoid the subtle assumptions involved in their application, alternatives are frequently proposed. Simple data transformations can be used to emphasize samples with multiple anomalies (i.e., samples that are anomalous in more than one element). For example, data values of several elements might be multiplied together. Each element may be ranked in a small number of classes (say 1-4 based on quartiles of the distribution), and the ranks added together. Analysis and display of these new variables then follows standard procedures.

Multivariate graphical methods (such as the shapes shown in Fig. 3.5) are used for grouping, although they do not necessarily lend themselves to automatic segregation. Methods for reducing the N-dimensional situation to a conventional scatter-plot format are another possibility. This approach includes the display of pairs of principal components and similar multivariate transforms. A useful technique developed by R. G. Garrett of the Geological Survey of Canada allows visual segregation of groups of samples, without requiring any statistical assumptions. The method is to determine the best-fit plane in the *N*-dimensional space defined by the variables, and then to plot a scatter diagram by projecting all samples onto this plane. Any distinct clusters are identified by drawing a polygon around them, and then removed from the data set (just like selecting a cluster of points on a scatter plot). Note that the cluster might be a single outlier so far removed from all other samples that they all appear to be in the same group. A new plane is fitted to the remaining samples, and a new plot is created. The process continues until the analyst decides to stop.

This process does not give any indication as to why the samples in each group are related. Similar to other visual methods, it simply identifies groups that appear to be related: The analyst then attempts to define the nature of the relationship. Some insight into the controlling factors may be gained by comparing the basic statistics of the groups. If we recall once again the aim of locating anomalous areas, it may be more important to see if the groups are spatially related. This determination is made by simply plotting all samples on a map, with a different symbol to represent each group.

The special graphical forms shown in Figure 3.5 may be used to visually segregate different groups of data. The first step is to choose the variables and the form that gives the best visual display (perhaps a polygon, multisided star, a tree, or a Chernoff face). The next step is to plot the shape for each sample, perhaps presenting them in a simple table for visual grouping. If a mapping program has the shape drawing routine integrated into it, the shapes may be drawn directly onto a map. Alternately, the groupings from a table may be entering as additional fields into the data file, and this field then would be used in mapping to choose different symbols on the plot.

Whole-Rock Analysis

When rock samples have been analyzed for all major rock-forming compounds, special multivariate methods may be used. Having the complete analysis puts a constraint on the data values; they must add up to 100 percent. This restriction can be exploited in comparing the amounts of each constituent to determine the type of rock. Visual determination of rock types is subjective, and rocks that are dissimilar in chemical and mineralogical terms may appear nearly identical. Proper identification of the rocks is essential to interpretation of the data, and in development of a geologic model of the mineralization processes. The advantage of a computer classification is that it is objective and unbiased. This is especially useful on large projects where several geologists are involved, since their subjective interpretations may not agree.

As for almost all other types of computer classification, the methods are not fool-proof. The results of the analysis are displayed in a variety of graphic formats for verification by the geologist. Various types of ternary diagrams are used to show the ratios of the major components, with boundaries drawn to show the distinct classes assigned by the program (Fig. 3.6). The assigned rock types may be plotted on a map to check the spatial relationships, and for comparison to existing geologic maps.

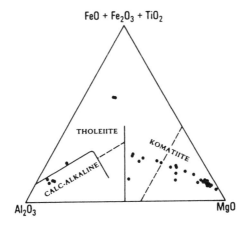

Figure 3.6 Ternary diagram for rock classification.
(Reprinted from *Precambrian Sulphide Deposits*, Geological Association of Canada Special Paper 25, 1982, p. 224, "Evidence for a volcanogenic-exhalative origin of a massive nickel sulphide deposit at Redstone, Timmins, Ontario", by D.J. Robinson and R.W. Hutchinson. Reproduced with permission of the Geological Association of Canada.)

CHAPTER 4

Geophysical
Data Analysis

Because there are so many types of geophysical data, this chapter examines a wide range of computer applications. Its broad scope is also a result of the long history of computers in geophysics, stemming from their essential role in handling vast quantities of data. With the exception of remote sensing (itself often considered a branch of geophysics), the other exploration techniques have adapted computers primarily to take over manual methods, while many geophysical techniques are essentially impossible without computers.

A perhaps natural result of this primary role of the computer is a greater acceptance of computer methods among geophysicists than in most other groups of Earth scientists. The nature of the job may be a factor: A field geophysicist often must adjust or repair electronic equipment, and thus may develop an affinity for working with such related devices as microcomputers.

Although many different exploration techniques are involved, the computer methods have many common features. The discussion under a specific method such as gravity may have considerable relevance to other types of geophysical data. All have similar requirements in data preparation, verification, and display. Ideally the same software should be applicable to these common tasks.

SEISMIC METHODS

Seismic exploration is today almost synonymous with oil exploration, since the technique is almost universally applied to locate potential oil-bearing structures. It is also synonymous with complex computing requirements, without which seismic reflection could not provide its detailed pictures of the structures in the Earth's crust. The details of seismic data processing are beyond the scope of this book, as the methods require highly specialized analysts. Only a brief review is given in this section.

Reflection Seismology

By far the most widely applied geophysical technique, at least in terms of expenditure, is seismic reflection surveying. Artificial sound waves (from explosives, vibrators, or other mechanical devices) are propagated into the Earth at many locations throughout the survey area (typically at fixed intervals along a series of parallel lines). A small portion of the sound-wave energy will be reflected at structural boundaries in the crust. The near vertical reflections are recorded with seismic detectors (geophones) at various surface locations. The responses of many geophones are recorded simultaneously on magnetic tapes (or other storage devices) for later processing. Figure 4.1 shows the method diagrammatically.

Reflection seismology is also used for detailed geotechnical studies. The basic procedures are the same, except that spacing of detectors, frequency sampling rates, and other parameters are adjusted to provide the higher resolution required.

The computer processing techniques are elaborate, usually requiring large computer systems and specialized people to produce results suitable for geologic interpretation. Initial data corrections employ many different methods from time-series analysis and signal processing, to compensate for noise, geometric distortions, and other effects. The primary result of this stage of processing is a seismic section, in which the data for each subsurface point are plotted as profiles (traces) side-by-side (Fig. 4.2).

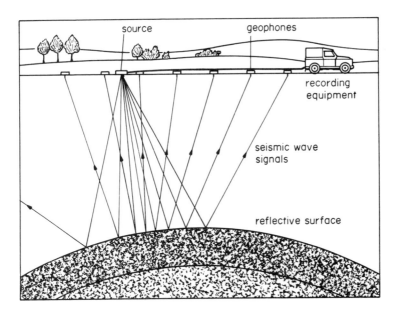

Figure 4.1 Basic principles of reflection seismology.
(Reprinted from *The Cambridge Encyclopedia of Earth Sciences*, p. 420, 1981, Cambridge University Press.)

Later analysis involves waveform modeling and other complex (and computer-intensive) procedures. As such, this topic is beyond the scope of this book, that is intended to cover a broad spectrum of methods which geoscientists may apply to different problems. For more details on the many types of computer applications used in this area, see the Bibliography.

The more general types of data analysis and display are widely used in reflection seismology, particularly in the interpretation stage. A major part of this process is to produce a series of maps of the structural boundaries in the crust, particularly those that might serve as traps for oil and gas. Contour maps, perspective views, and color maps are the typical forms of displaying the structural surfaces (combinations of these can be highly effective). To produce maps, the interpreter must identify structural horizons on seismic cross-sections (or other forms of seismic data display), convert these "picks" into *X-Y-Z* positions by attaching a coordinate reference, and then use a general-purpose mapping program to create the display. Computer software may aid in making the picks,

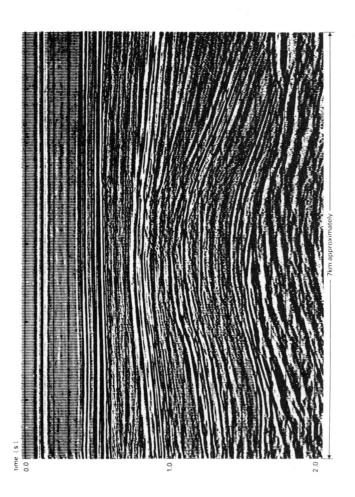

Figure 4.2 A typical seismic section.

The horizontal axis represents distance along the survey line. The vertical axis is time—the measured delay between creating seismic waves at the surface and the waves returning from reflection off subsurface boundaries. (Reprinted from *The Cambridge Encyclopedia of Earth Sciences*, p. 421, 1981, Cambridge University Press.)

by scanning for characteristic waveforms on adjacent seismic traces. Once the picks are made, a general-purpose mapping program can be used to display the structural surfaces.

Special software may be applied in interpretation, for example, to migrate the seismic picks from time to depth positions. The basic forms of seismic data display use travel time of the reflected waves as the vertical coordinate, not taking into account varying velocities, nonvertical travel paths, and a variety of other effects. Migration attempts to correct for this simplification using a layered velocity model. It may be applied during the normal processing sequence, or applied after the interpreter has made the picks to be used in mapping. Modeling techniques are used to produce pseudo-lithologic sections, in which the characteristics of the reflected waveforms are analyzed to determine the types of rocks causing the reflections.

Refraction Surveys

Seismic methods also include refraction surveys, which are not as powerful in mapping subsurface structure, but nevertheless are of great value in near-surface studies. The much simpler procedures mean lower costs, which can be an important consideration in regional studies or for other applications that do not have the same potential for a large payback as does oil exploration.

The refraction method is aimed at determining near-surface layering, by measuring the speed of seismic waves traveling through the near surface layers. A basic assumption is that the characteristic velocities for seismic waves are different in each layer, so that waves that travel primarily through deeper layers will arrive at distant sensors (geophones) at a different time than those that remain confined to the shallower zones. In general, seismic waves travel faster in deeper layers (due to higher pressures and greater compaction of the rocks), so that the longer path length below the surface layers and along a deeper layer can result in a shorter travel time. By determining the horizontal distance at which waves from deeper zones become the first arrivals at a given detector, it is possible to calculate the depth to the interface between the layers. The process is illustrated in Figure 4.3.

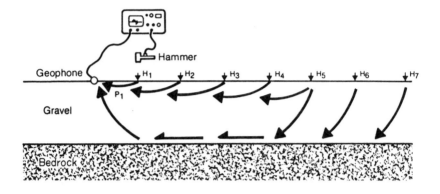

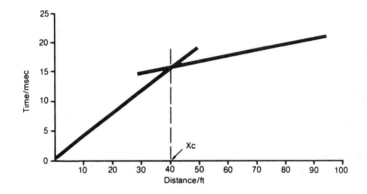

Figure 4.3 Basic principle of refraction seismology. (Reprinted from *Practical Geophysics for the Exploration Geologist*, Northwest Mining Association, 1980, p. 166, chap. 4, "Applications for shallow exploration seismographs," by D.B. Crice.)

The involvement of a computer may occur at several stages of the process. Given the small amount of data and straightforward computational methods, small computers are more than adequate for the tasks. Typically the response of the geophones will be recorded by a specialized computer, which will also track elapsed time after the shot. Picking the arrival time of the first seismic wave at each geophone is the first stage of analysis. This is not always a simple task because of spurious signals from wind and other uncontrollable sources, so that the picks made by a program may have to be manually edited. Once the picks are made, the calculation of velocities and depths is relatively straightforward. This step is also dependent on an assumed layering model (i.e., how many layers, whether they are horizontal and dipping), so once again the results from the computer must be verified by the geophysicist or technician doing the interpretation.

Besides the calculation of velocities and depths, the computer may be used to draw a cross-sectional view of the results. The form of display also depends on the type of layering model being used. Frequently other types of data may also be available, so the graphical display of the refraction results may be only part of the computer plotting. For example, the surface topographic profile and other geophysical measurements such as magnetics and gravity are often shown along with the subsurface structure. In such cases, a general-purpose section drawing program has many advantages over one that is specifically attuned to the refraction problem. This comment can be made in many other situations as well, as we have noted elsewhere in the book.

GRAVITY AND MAGNETICS

Although they measure different properties of the Earth, gravity and magnetic surveys are usually considered together as potential field methods. The term stems from the mathematical formulation of the potential, which is used to calculate gravitational and magnetic fields of geologic models. The same theoretical framework applies to both, except that gravity is a monopole field (i.e., a single-point mass is the basis of all calculations), whereas magnetics stem from dipoles (considered as an infinitesimal line element that has a direction as well as a unit of intensity). Stated differently, the gravitational potential is a scalar quantity, while magnetic potential is a vector.

Data Reduction

The first problem that arises with gravity and magnetic surveys is that the measured fields require considerable manipulation before actual analysis of the data can begin. This stage is called data reduction, as distinct from later analysis, interpretation, and modeling steps.

Reduction corrects for instrumental problems as well as basic characteristics of the Earth's gravity and magnetic fields. The aim is to reduce the
raw observations to those components due to near-surface structures,
since the object of exploration is to determine the characteristics of this
part of the crust. Gravity and magnetic data can also provide useful
information on the deeper parts of the Earth (beyond the reach of
extraction techniques). The procedures used for whole-earth studies are
similar to those applied in exploration, but are designed to extract
different components of the total field.

Drift of the measurements is the first correction to be made. In
gravity surveys, drift is largely due to unavoidable changes in the
mechanical behavior of the instrument (temperature expansion of springs,
for example). To correct for drift, surveys are done in loops, with
repeated measurements at the same point on the Earth's surface at
different times. Assuming the drift has been linear (i.e., proportional to
time elapsed after the first reading), all values taken between the
repeats can be adjusted by prorating the difference measured at the
repeat station. For example, if the reading at the base station has
changed by 5 milligals after two hours, the station recorded one hour
after the first base-station reading would be adjusted by 2.5 milligals.
Although the calculations are relatively easy to do manually, it is
obviously a fairly simple task to write a program to do the job automatically. All that is required is that a time reference be recorded with each
gravity observation, and that the repeat points are noted in some way
that can be detected by the program.

For magnetic data, the problem is similar, but the cause of drift is
mostly the dynamic field itself (i.e., the Earth's magnetic field changes
over short periods of time). The looping procedure can also be applied for
magnetics, provided that the return time is short enough for the
assumption of linear change to be valid. A better solution is to use two
magnetometers, one doing the survey and another fixed in one place
simply recording the changes in the field. With a time reference attached
to each set of readings, corrections can be made to the survey readings
directly from the observed change at the fixed base station. Note that
this method no longer requires the assumption of linearity in time, but
does assume that the maximum distance of any survey point from the
base station is small compared with the scale of spatial variations in the

field. In other words, it is assumed that the magnetic field variations are the same throughout the survey area.

Whichever method is used, it is again an easy programming task to apply the corrections. Most magnetic surveys are now done with digital magnetometers, that store readings in internal memory for later offloading into a small computer. A time stamp is included with each reading. Assuming that the clock in the base-station instrument and the survey instrument are synchronized, the drift correction is a simple matter of matching times. Note that some interpolation may be required between adjacent base station readings. The time increment at the base station therefore must be set short enough so that the field change can be assumed to be linear in that time period. During magnetic storms, this period might be short indeed, so that even with a base-station recording, the accuracy of the correction cannot be guaranteed. Normal practice is to suspend magnetic surveying during such periods.

Besides applying drift corrections, a computer used in data collection can perform various tests for validity of the data. Tests are especially important with automatic recording devices, where the operator may have little control. A disturbing trend in geophysical surveys is that the operators do not always have the level of training and experience that was the norm in the past. The result may be that if the system is malfunctioning, the operator does not detect the problem until many erroneous data have been recorded. To perform validity checks, some assumptions about maximum rates of change between readings are incorporated into the software. Depending on the perceived severity of detected errors, data may be rejected outright, or flagged in the computer file for later manual checking. Such programs often produce graphical displays (such as profile plots along each survey line) to ease the chore of manual verification.

After recording and verifying the basic observations, corrections are made to remove the whole-earth effects (which are the major component of the field, especially for gravity). The reduction procedure depends on theoretical models to predict the gravitational and magnetic fields at each survey point. For gravity, these can be expressed as simple formulae related to the latitude and elevation of the station. Accurate location data (surface coordinates and elevations) are essential, since the magnitude of the corrections is generally much larger the gravitational effect of near-surface sources. The objective of the survey is to detect such small anomalies in the observed field.

For magnetics, a model of the Earth's field is needed, expressed as coefficients of spherical harmonics (up to the tenth or twelfth order). The predicted value of the field depends on these coefficients (literally hundreds of them), the location of the station, and the time of the reading. These calculations are only practical using a computer, unlike the simpler gravitational calculations that can be done manually. For surveys confined to a small area, the correction is often omitted, since the computed values will be essentially constant over the survey area, or at worst imply a minor linear gradient to the readings.

Because the Earth's magnetic field changes over time, the model must take into account the expected change since the model was developed. Coefficients for the secular variation (annual changes in the field) are included in the model. The date of the survey readings is thus needed to compute the theoretical magnetic field. Magnetic models are revised periodically on the basis of new observations from satellites, so a magnetic correction program must also be revised from time to time (this is a simple matter of replacing the coefficients, which can be obtained in computer format as well as in published reports).

In areas of significant topographic variation, the surrounding terrain can disturb the gravity field. These corrections are considerably more complex, requiring a model of the terrain in the area. For each survey point, the effect of hills and valleys (i.e., departures from a level plain) can be calculated using the theoretical formulae for simple shapes (usually rectangular prisms corresponding to a gridded model of topography). Some assumptions about density changes are required, although they may not be critical if the geology is fairly uniform (i.e., the entire region is composed of similar types of rocks). The computer implementation is to search for all cells in the grid within a specified maximum distance of the station. The gravity effect of a prism is then calculated, using the difference between the grid elevation and the station elevation as the height of the prism. The availability of digital terrain models (DTM) is usually a limiting factor, although it is becoming less so as government mapping agencies become more computerized. Terrain effects can also be important in magnetics, but may be even more difficult to correct, since the magnetic susceptibility can vary more rapidly than density within the surface rocks.

Common problems in large surveys are misties, or conflicting data values where survey lines cross. They can arise from many sources, for example, errors in drift corrections, differences in ambient noise levels at different times, errors in location, and so on. In airborne surveys, crossing lines may be flown at different altitudes, which can result in misties due to errors or uncertainty in the leveling procedure used to correct for altitude differences.

If misties are present, contour maps will show many apparent anomalies at line intersections. To remove (or at least minimize) these effects, values are distributed along each line, in a procedure similar to drift corrections. The problem may be considerably more complex, however, since there may be hundreds (or thousands) of intersections to be checked, and the adjustments are intended to resolve all of the differences.

Display of Potential Field Data

After initial corrections, the next step is to examine the data for regional trends and local anomalies. Large volumes of data are typical, especially with airborne surveys, so the only practical way to do this is with computer-generated plots. As already noted, profile plots may be generated in the data acquisition stage, but these are not sufficient when dealing with multiple survey lines. Maps showing all of the data are the most common type of data presentation, although they may come in a variety of forms. Individual geophysicists may prefer different types of maps, and often more than one type may be created, as each has its own advantages.

Figure 4.4 shows a stacked profile map, combining simple profile plots for various survey lines onto a map projection. This format shows all values as recorded, but can be confusing when the survey lines run in several directions. An interactive approach may be useful in creating the map, because different profile scaling factors can dramatically change the appearance.

Figure 4.5 shows a portion of the same data in a posted data format (the practice of writing numbers on a map is called *posting*). This format gives the most immediate presentation of actual values, since they are read directly, and do not have to be inferred from the offset of the profile.

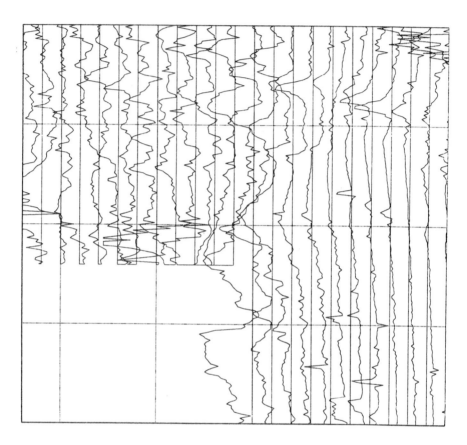

Figure 4.4 Stacked profile map.
Ground magnetic readings are represented by the offset of the profile from the base line, which is plotted along the location of the survey points.

It requires detailed examination to detect anomalies, however, and also is prone to problems of overposting. At the desired scale, adjacent readings may appear so close together that it is impossible to fit all of the numeric values on the plot. Posted maps are used only when the data are to be contoured manually, or as a check on a computer contoured map.

A contour map is useful for marking trends that cross several lines (Fig. 4.6). Computer contouring requires considerable manipulation of the basic data, which can result in misleading trends or anomalies that

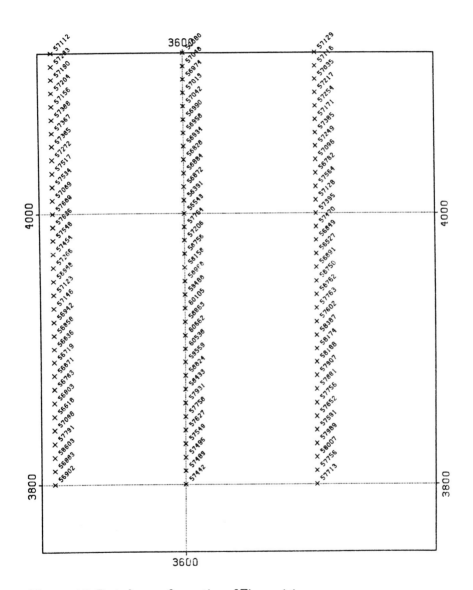

Figure 4.5 Posted map of a portion of Figure 4.4.
Note that the actual data values now can be read directly, but the location of anomalies requires careful examination of the numbers.

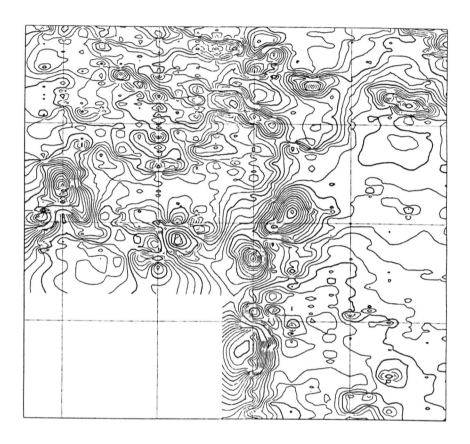

Figure 4.6 Contour map corresponding to Figure 4.4.
Regional trends are more evident, but details along each line are obscured.

are mainly artifacts of the contouring process, and not representative of the real Earth. As noted in Chapter 2, contour programs work in two stages: first interpolating the raw data onto an organized coordinate system (i.e., gridding or triangulation), then tracing lines through the interpolated data.

The revised coordinates are most often a square or rectangular grid (i.e., values at regular intervals along the X and Y axes of the mapping grid). If the survey locations are already in a rectangular system, this procedure may be relatively straightforward. However, this is frequently not the case, so that the mathematical technique for estimating the data values at each grid corner must be very sophisticated to avoid unrealistic estimations.

Some gridding programs take advantage of potential field theory, and use the constraints implied by the potential field equations in interpolation between data points. Since the data are often acquired along lines, this type of data distribution may also be assumed. Gridding techniques that assume essentially random data distribution can have problems with line data, because the sample spacing (i.e., distance between readings along each line) is typically much less than the distance between lines. More stable results are possible by exploiting this fact, first by interpolating along each line and then between the lines. Cross-correlation may be used to follow trends between lines. The user may define a trend direction based on prior geologic knowledge. The gridding algorithm biases its search for data points in this direction, which enables a smoother transition across gaps in the line coverage.

Data Enhancement Techniques

The features of exploration interest are not always readily visible in displays of the basic data. Strong regional trends, superposition of multiple anomalies, distortions inherent to potential fields, and other obscuring effects can make interpretation of the data nearly impossible. A variety of mathematical operations may be used to remove or at least reduce these problems.

Many techniques take the empirical view that the anomalies of interest are a signal that is not measured directly, since the observed data are a mixture of this signal and various unspecified sources of noise. The noise might be the traditional type, unavoidable errors in the measurement system, or it may be caused by real sources that are not of interest. Even though the nature of the noise may not be known, it is necessary to make some assumptions about its characteristics to develop methods that distinguish noise from signal.

For random noise (i.e., uncontrollable errors which may occur in any measurement), simple averaging of data values is the most common method of noise reduction. This means that multiple values of the basic data are averaged together, either by taking several readings at each location, or by averaging adjacent locations. It is obviously necessary to anticipate the need for averaging (by a small trial survey, perhaps), so that sufficient readings are made to allow for later averaging. Averaging may suppress more than the random noise, and thus can mask small real features in the data. When the objective is to find large regional components however, this suppression is not a problem, and the simplicity of the method is in itself an advantage. An averaging algorithm is easy to write, and is relatively undemanding of computer resources.

To avoid the masking of local features, more sophisticated techniques are required. One approach is to view noise removal as a filtering problem. Random errors at each station then will appear as high-frequency components (theoretically at the highest frequency that can be represented with the particular spacing of survey points). An alternate description is that lower frequency components are trends that can be detected across multiple adjacent stations. Filtering operators can be designed with a fairly sharp cutoff to remove such high frequencies, while leaving lower frequencies essentially untouched. Note that local averaging is just a special case of the more general spatial filtering procedure, but one that does not provide the desirable sharp cutoff.

The choice of filter depends on the size of features to be retained. In addition to the high-cut filter just described, it might be desirable to also remove large trends (i.e., low frequencies). Both operations may also be combined in a band-pass filter. In practice it is impossible to attain perfectly sharp cutoffs in filtering, due to the finite number of data available and computational limits on the number of points in the filter operator. The result is that some of the desired frequencies may be partially suppressed, and the other frequencies are not completely removed. An alternate way to get the desired result may be to subtract the results of filtering from the original data, for example, using a low-pass filter (or heavy smoothing) to enhance regional trends, and to subtract the filtered values from the total response to emphasize smaller features.

Filtering can also be done directly in terms of frequency components, by adjusting values of the frequency components of the data. Although this frequency domain method may appear simpler than the corresponding convolution of a filter operator with the original data (in the spatial domain), it does not avoid the fundamental limitations noted above. Applying cutoffs in the frequency domain must allow for smooth transitions from the accepted region to the rejected frequencies, to avoid generation of spurious features on transforming the data back into the spatial domain.

A filtering program should allow for several types of filters to be used. As in similar data transformation procedures, the output should be in a form suitable for use in other programs. Note also that filtering may be applied to original data or to a gridded representation. If the data spacing is not regular, however, the algorithms become much more complex, and many filtering programs will work only on gridded data (especially when working in two dimensions). The result may be unwanted enhancement of artifacts in the data introduced by gridding problems.

A variety of image processing methods are useful for enhancement of geophysical data, including gravity and magnetics. These are briefly discussed in Chapter 5.

Aids to Interpretation

Filtering or smoothing may be applied to all types of data. There are also mathematical operations specific to potential field data that are used for purposes similar to filtering. The first (and probably most widely used) of these is continuation, which transforms the data to the form they would take if the survey was done at a different elevation.

Since potential fields decline in strength with distance from their source, increasing the elevation of a survey (e.g., by flying at a higher elevation) changes the response observed at each station. The effect is variable, however, being strongest for stations vertically above the source, and negligible for distant survey points (the change in the vertical dimension makes only a small contribution to the total distance from the source). The suppressing effect is also greater for near-surface sources than for deeper ones, since the change in elevation is a greater proportion of the total distance. The end result is that the potential field anomaly due to a deeper source is broader than for a shallow one.

While taking readings at higher elevations is possible for airborne surveys, it is obviously not practical for ground surveys. Even with airborne work, it is not practical to redo the survey at different elevations to compare the responses. Fortunately, relatively simple algorithms to perform upward continuation can be derived from the potential field equations. As for wavelength filtering, the operation may be performed in the spatial or frequency domains, with similar concerns for size of operators, frequency tapering effects, and so on. The effect of upward continuation is to suppress local anomalies (i.e., near-surface sources) much more than the more deeply-seated regional effects (Fig. 4.7A and B). While this procedure enhances the regional field, we must remember that enhancement is achieved by relatively less suppression. The field is still reduced, so the simple subtraction method for deriving local anomalies cannot be used. It may be possible with simplifying assumptions and scaling factors to produce something analogous to a local feature map, but it is dangerous to carry this procedure too far.

As easily as moving away from a source, the potential field equations can be manipulated to create the downward continuation function. This operation should have the opposite effect: All features are enhanced, but the near-surface ones more than the deeper sources. In practice, however, downward continuation is a considerably more difficult problem. The amplifying effect applies to all components of the data, including any random noise that may be present. Since random noise looks like near-surface sources, it receives the greatest boost from continuation, and can quickly overwhelm all other features.

Figure 4.7C shows the result of continuing downward, using the same data. Note that the continuation distance is half that used for upward continuation in Figure 4.7B, but the effect on the data is much greater. (Downward continuing the same distance as in Figure 4.7B produced an essentially unreadable map.)

For magnetic data, the variable inclination of the magnetic field is a complication. Anomalies due to simple source bodies take on a fairly complex appearance as a result. The characteristic shape varies with the orientation of the Earth's field (and thus with location on the Earth's surface). Another transformation of apparent position can help here: Reduction to the pole produces a magnetic field response that would be observed if the survey (and sources) were at the magnetic pole. Since the magnetic field is vertical at the pole, the effect is to produce simple shapes similar to the gravity response of the same body. An extension of the technique, called *pseudogravity*, attempts to create an equivalent field, by vertically integrating the magnetic response to reduce it to a monopole field.

Figure 4.7 Continuation of potential fields.
A: Map of gravity data, from surface readings.
B: Same data as in A, after upward continuation of one grid spacing. Note the smoother appearance of the contours.
C: Gravity data downward continued one-half grid spacing.

All of these operations can be considered as variations on the basic filtering procedure. They may be performed in one dimension (i.e., along individual survey lines), or two dimensions (at regular grid points over the survey area). In either case, it may be necessary to interpolate regularly spaced values from original data locations. Similar to filtering routines, geophysical transformation algorithms are written typically assuming that only gridded data will be input, to allow use of matrix methods which greatly reduce computation time. Also like filtering, problems in gridding may be magnified after further operations.

It should be noted that applying these techniques requires some assumptions about the type of body that produced the anomaly. For one-dimensional data, the potential field equations are based on two-dimensional bodies, with infinite extent in the third dimension. This assumption implies that the gravity/magnetic profiles are taken along a line perpendicular to the long axis of the body, and that its length is much greater than its width.

The key to success of these methods is to have them integrated with a complete data analysis system, including statistical functions for comparing derived and original data, and easy-to-use display programs to plot the data on maps and in other forms. In general, it is better to have a separate data display program with a wide range of options, than to directly link such functions into each analysis routine. Some graphics options are highly desirable in analysis, of course: It can become tedious to keep jumping out of one program into another whenever graphics are required. The trick is to obtain a good compromise between having too many separate modules, and having many duplicated functions. Where duplication exists, the method of use should be identical. Consistency leads to more efficient use of all parts of a large system, in that the experience gained in one part is at least partly applicable to others.

These principles are not just for potential field data, or even restricted to geophysical surveys. At the risk of being repetitious, I must emphasize the value of applying common techniques to all types of exploration data.

Modeling and Interpretation

The aim of the various display and enhancement methods is to allow an interpreter to identify anomalies in the potential field observations. The next step is to relate these anomalies to geologic parameters of the Earth's crust, and to use them to add to the knowledge of the survey area. Computer algorithms are also used in this stage, to help define the possible geologic models that are consistent with the data.

Defining an acceptable model may be achieved in one of two ways. First, a geologic model may be defined as a simplified description of the structural features of the crust that may be causing the anomaly. For certain geometric configurations, the expected gravity or magnetic response of the model can be computed. This is called solving the forward problem (or direct problem). The form of the model might be a set of simple shapes (prisms or spheres, for example), an arbitrary shape defined as a set of polygonal boundaries, or a specific geological feature (a contact between rock types, or a vertical dike, for example).

By comparing the theoretical response to the observed data, it may be possible to adjust the parameters of the model to produce a better fit. The alternate approach is to solve the inverse problem, that is, to compute values for geologic parameters from the data. In general this method is considerably more difficult, both in mathematical derivation and in practical use.

Since there are an infinite number of solutions to the inverse problem for potential field data, it is essential to be able to easily compare the results from different geologic models. This is equally true regardless of the approach taken. The requirement of looking at different results may not seem obvious when using an inversion method, which may appear to give a single answer to the problem. Because of the nonuniqueness, the judgment and experience of the interpreter must always be a major part of selecting the final model. Another aspect of nonuniqueness is that restrictive assumptions must be built into the modeling process. For example, it may be assumed that the source of a gravity anomaly has constant density, so that the observed anomaly is due strictly to geometric parameters (shape, position, depth of burial, etc.). Since the assumptions may be invalid, it is also useful for a modeling program to allow variations in the type of model to be tested.

RADIOMETRIC METHODS

Many of the basic requirements for computer analysis of gravity and magnetics also apply to other types of geophysical measurements. In other situations, the governing physics may not be as easily quantified, so that the emphasis is more on basic data display and anomaly recognition than on modeling and inversion.

Radiometric surveys are aimed at identifying particular types of rocks or minerals whose trace components include radioactive isotopes. Radiation in the form of gamma rays and X-rays is measured at fixed intervals along survey lines. Detectors typically respond to several energy levels corresponding to (relatively) abundant isotopes of potassium, uranium and thorium. The total gamma-ray response is taken as well. Although there is some overlap in the energy levels, it is possible to estimate the amounts of the radioactive isotopes in the near-surface rocks with simple mathematical formulae. Corrections for survey elevation are also required in airborne surveys. The procedure to apply corrections and calculate potassium-uranium-thorium values is fairly automatic, done as part of data collection. The raw data from a radiometric survey are often in the final form to be used in generating statistics and graphical displays.

Since finding anomalous areas is the main objective, the main demand in computer analysis is to produce effective displays of the data. The same methods used for other types of data are equally effective here. Simple data postings are perhaps less common, simply because the number of data is typically large. Radiometrics are often used in regional exploration, typically from aircraft that can cover large areas and collect many thousands of readings in a short time.

The data spacing along lines is typically much smaller than between lines, which can cause problems in gridding and contouring routines. Averaging data along lines is a simple procedure that frequently reduces such difficulties. It can also allow considerable savings in computer time. The averaging calculation is simpler than the areal searching and interpolation schemes used in contouring, and typically results in a much smaller data set. As the execution time is strongly dependent on the number of data values, contouring may be completely impractical on the original data set, which might be ten or twenty times larger than the averaged version.

Radiometric anomalies may be more evident in ratios of the components. A ratio may amplify relatively small variations in each value, to provide greater contrast with background areas. A general-purpose numerical operation program is therefore an important part of a geophysical analysis system. It should allow adding computed quantities as new fields in the data file, to permit statistical comparisons with original data. Plotting observed and derived quantities on the same map is then a simple matter as well.

Correlations between the different components can also be useful. The objective here is to locate joint anomalies, which may be done individually for each survey line, or in a two-dimensional sense over an area. Correlation functions can be computed to quantify the relationships. For the two-dimensional case, this computation requires a regular grid representation of each variable. Contouring programs based on grids normally allow grids to be saved, so this step may not have to repeated for the correlation.

More often the comparison is done visually by plotting more than one variable on the same map (or profile plot, for the one-dimensional case). The trick is to find a display format that makes it easy to distinguish each variable. This might involve two sets of contours in different colors or line styles, or showing one variable with contours and the other with scaled symbols. In other words, use two different methods for highlighting anomalies in a single variable on the same plot. Since the procedure may be repeated on other data sets, the software should allow for saving the display parameters, and for using such a parameter file as an optional replacement for interactive parameter setting.

ELECTRICAL METHODS

There are many geophysical techniques for measuring the electrical characteristics of the crustal rocks. In exploration, they are used to find various types of minerals that are good electrical conductors (at least compared to the common rocks). They may also be applied in exploration for fluids that may alter the electrical properties of the rocks; for example, water in porous rocks may make them highly conductive. In petroleum exploration and development, the contrast between zones containing water and hydrocarbons may be readily detected by electrical measurements.

Electrical methods are classified as active techniques, in that the survey involves stimulating a response from the Earth by transmitting currents (or other forms of energy) into the ground. In this sense they are similar to seismic methods, and distinct from passive techniques such as gravity and magnetic surveys, which simply measure properties of the rocks that are always present. There are many different ways to transmit currents into the ground, and many ways in which the Earth may respond to this energy. The great variety of survey methods that result need not be discussed in detail here, as the concern is with the computer treatment of the observations.

As with the other geophysical methods, the first involvement with the computer is often in data collection. Instrument responses may be recorded automatically in memory, and later dumped into a small computer for preliminary analysis. For surface surveys, this procedure might be plotting profiles for each survey line, just as is done for other types of data.

Some methods (such as induced polarization [IP]), rely on multiple measurements using different source-receiver configurations to achieve different levels of depth penetration. These methods are analogous to seismic refraction, in which greater source-receiver offsets detect seismic waves from deeper layers. The basic data display for IP data is a pseudosection, in which results from larger spacings are plotted at apparently deeper positions (Fig. 4.8). For field work, these may be plotted on a printer. Computer contouring may be applied to each pseudosection. For large surveys, this step may be left until all data are available for a combined display on a large plotter.

Many geophysical measurements are taken in drillholes. The basic principle of data display is similar to surface work: Show the responses as profiles along the drillhole. This problem may be more complicated, in that the drillhole may follow a deviated path that must be translated into X-Y-Z coordinates prior to plotting. A data merging operation is then required, since the positional information is normally acquired independently of the geophysical measurements. The link between the two types of data is depth along the drillhole. Some assumptions about the actual path of the hole are needed, since the location survey is taken

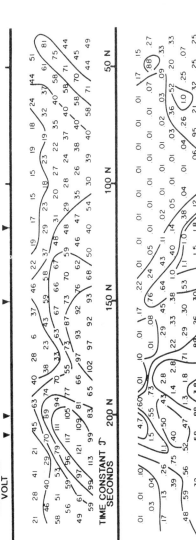

Figure 4.8 Pseudosection display for IP data.
Each level of plotted values represents readings at the same source-receiver geometry. Lower levels are for greater separations, which correspond to the electrical response of deeper layers in the Earth. (Reprinted from L. E. Reed, 1989, "Geophysics in Gold Exploration—Some Examples," Fig. 37.9, p. 481, in *Proceedings of Exploration 87*, Ontario Geological Survey, Special Volume 3, pp. 473–485.)

at selected points along the hole, which are not necessarily the significant changes in direction. For multiple drillholes, these calculations are combined with a plotting program to produce a sectional view with the holes in proper perspective. This view is similar to a stacked profile map, except that the view is not of the Earth's surface. More details on analysis and display of drillhole data will follow in Chapter 6. The display forms illustrated there can be applied to geophysical data as well as to other types.

ELECTROMAGNETIC METHODS

Another category of active methods electromagnetic (EM) techniques, which involve recording the Earth's response to transmitted electromagnetic waves. The range of possibilities is virtually infinite, since different portions of the broad electromagnetic spectrum may have different effects. Electromagnetic techniques are similar to the electrical techniques in that the primary objective is to detect (relatively) conductive rocks. The penetration characteristics may be very different, however.

One widely used method appears at first to be a passive technique. Very low frequency (VLF) surveys are taken with only a receiver to detect the electromagnetic energy; no source is required. The energy source is the carrier wave used for worldwide communication with American and Russian military submarines. While the information transmitted by the land stations is not of interest to the geophysicist (and obviously is not interpretable in any event), the source location and frequency is public knowledge. Since the transmissions are intended to travel great distances, most places on the Earth receive (and respond to) some of this energy.

One problem can be that the alignment of source and target has a major effect on the measured response. Conductors in line with the magnetic fields generated by the source will generate little response. In many areas, more than one transmitting station may be usable, so that different geometries may be used to avoid this problem. The great savings in survey cost by not having to deploy both transmitters and receivers more than compensates for this difficulty.

Computer treatment of VLF data is similar to treatment of gravity and magnetic data. Readings along lines must be stored, along with positional information. Coordinates may be attached to each reading manually, or by interpolating along the lines from a few digitized points. Data analysis is often a matter of plotting profiles along lines, or stacked profiles or contours on maps, and looking for "kicks" in the readings that indicate contacts between conductive and resistive rocks. The transitions in the data may be enhanced by filtering, for example the Fraser filter that calculates the difference between adjacent pairs of values. Like many of the other methods, the physics of the electromagnetic response are not as well understood and quantified as for potential fields, so the modeling approach to interpretation is not as common.

Some EM methods are truly passive in the sense that they do not use an artificial source of energy. Instead, the Earth's response to natural electromagnetic waves is measured. Magnetotellurics (MT) measure the Earth's response to variations in the magnetic field, for example. Data acquisition and processing are similar to VLF.

GROUND-PROBING RADAR

Radar might be considered an EM method, but is usually considered as a separate field. In essence, it has the same strategy as reflection seismology: to detect structural boundaries by timing the returns of waves transmitted from the surface and reflected from subsurface boundaries. Due to the greater velocities of radio waves, time intervals are extremely short, and the required electronics have been developed only recently.

Computer processing is also similar to seismic data procedures, generally being a job for specialists. General-purpose mapping programs may be used in the interpretation stage: The main difference is simply the scale (since radar is used to map smaller features than seismic surveys).

Applications are restricted to near-surface problems, since absorption and scattering of the radio waves limits depth penetration. Highly detailed geotechnical investigations, mapping below overburden, checking for structural problems in roads and bridges, and measuring snow and ice thickness are some of the problems to which radar provides a unique solution. Since radar reflectivity is highly dependent on water content, its depth penetration may be limited. On the other hand, this limitation is an advantage in exploring for groundwater and other hydrological applications.

CHAPTER 5

Remote Sensing and Image Analysis

In a broad sense, exploration techniques such as geochemical and geophysical surveys are a form of remote sensing, in that they attempt to determine geological characteristics of the Earth without requiring direct contact with the target region. Normally, however, remote sensing has a more restricted meaning: spectral scanning systems operated from satellites or aircraft. The application of these methods is widespread, including geological studies, agriculture, forestry, and meteorology. The military aspects should not be forgotten as well; remote-sensing systems would not have their present level of sophistication without the major research and development role taken by governments interested in advanced surveillance techniques.

Image analysis is almost synonymous with remote sensing. Satellite spectral scanners generate essentially the same type of data as photographic scanners, and the various radiation scanners used in medicine. These fields are also heavily dependent on image processing.

In the context of this discussion, an *image* is a representation of visual data by arrays of discrete points. Data of this type are said to be in raster format (as opposed to vector format, in which a picture is represented by a set of lines). Compare a newspaper photograph, which is composed of many dots, to a line drawing of a building. *Visual* has a special connotation as well; the data are displayed graphically, although

105

the measured quantities may be completely invisible to the human eye. At each point (called a picture element, or pixel), spectral components are defined by one or more numbers. The more resolution desired in the spectrum, the more values that must be measured at each pixel. The LANDSAT satellites, for example, take readings for seven discrete bands in the electromagnetic spectrum, ranging from the visible region into the infrared.

Image analysis may be defined loosely as the set of mathematical and graphical operations used to transform such data into interpretable pictures. Although the main applications are with data acquired directly in raster format, the techniques are being used for other types of data. For example, conventional photographs may be represented in raster form using scanning devices, so that image analysis may be applied to improve resolution or extract subtle features. Scanning is increasingly used to digitize many types of drawings and documents. The objective may be simply to convert the data for redisplay at different scales, or to prepare them for computer analysis. Geochemical or geophysical data may be interpolated onto a regular grid, and then visually enhanced using image analysis software.

Like reflection seismology (which was briefly discussed in the previous chapter), much of the computer involvement in remote sensing and image analysis is at an advanced level. Powerful central processing units (CPUs), high-resolution color graphics, sophisticated software, and highly trained analysts are common components of these systems. At present, they are virtually essential components as well, so that most of the computer work is not done by the end user of the data. With increasing power in smaller machines, and more frequent application of the methods, the situation is changing. It is now fairly practical to use a small computer for interactive display of pre-processed images. The complex corrections and sophisticated analysis techniques are done by specialists, but the results may be delivered to the end user in a form suitable for viewing with a personal computer. Display programs used on small computers have many options for assigning colors, zooming in and out, and doing some simple transformations.

In the near future it will be more common to have remote-sensing capabilities in small computer systems, and to see many scientists using remotely sensed data. Image-analysis methods will also be more routinely applied to other types of exploration data. Whether the complete range of image-processing functions are applied by end users, or remain in the hands of specialists is difficult to predict.

This chapter is a brief overview of the field, concentrating on how the procedures relate to other computer applications in exploration. Once again, the objective is to discuss methods that are in relatively common use, or can be fairly easily implemented by nonspecialists. See the Bibliography for references that give detailed descriptions of the more advanced computing techniques.

DATA PREPARATION

Like many of the examples in earlier chapters, processing of image data may be considered in two phases. Various corrections and transformations must be applied to the raw data prior to the actual analysis. Some of these are fairly general, and are needed for many types of images, while others are specific to the spectral characteristics of the data. Most of the basic corrections are applied as part of data acquisition; that is, they have already been done when the user receives the data. (Due to the complex and expensive equipment, virtually all remote sensing is done by specialist agencies or companies, and is not undertaken by the organizations that will use the data). Although the end user must assume that these operations were successful, it is important to have a basic understanding of the procedures to recognize potential problems that may appear later.

Satellite sensors are similar to other pieces of electronic equipment in that they do not always work exactly to specifications. Since they cannot be repaired easily, it is necessary to accept some problems in the basic data, and attempt to correct them as much as possible.

The first stage of processing is to scan for missing or abnormal data along the path of the satellite. Most sensor systems have an array of detectors oscillating back and forth across the track of the satellite, to give a wider range of view on the ground. Problems in the sensor itself, or in transmission of the data to the ground, may lead to some scan lines occasionally being lost, which would appear as horizontal black or white lines on the image. To correct for such missing data, values are interpolated from the adjacent lines. Obviously there are limits to this method: If several adjacent lines are missing, the interpolation distance would be too great to expect a reasonable result.

The sensor system is composed of multiple detectors, so that one scan generates several lines of pixels. If the detectors are not perfectly matched, their output levels will not match, causing abrupt shifts in value between adjacent pixels. These shifts will appear as a repeating pattern N pixels wide, where N is the number of detectors in the sensor. This problem is removed by destriping, which computes the average statistics of the signal level from each detector (by selecting every Nth line in turn), and adjusts all lines for each detector to the overall average.

The last routine correction is needed to compensate for distortion of the pixels. The oscillating sensor receives data at different angles from the ground, so that the effective area is not uniform. In addition, forward motion during a scan imparts a skew to the line. The process of geometric correction is designed to remove these effects. It involves a complete interpolation of the original pixel values onto a new set of coordinates, that have a fixed spacing on the ground. This is a standard gridding problem, although it involves much more data that is typical for other applications. Because of the density and approximate regularity of the input data, the gridding algorithms can be simplified for speed, and are more stable than for those designed for randomly located data.

Note, however, that geometric correction does not produce coordinates registered to a particular map projection. This requires yet another interpolation onto a new grid coordinate system such as Universal Transverse Mercator (UTM, commonly used for topographic maps). This process is called registration or geocoding, and is not part of standard preprocessing of satellite images. It is also not a technique practical for small computers, at least when dealing with full scenes from a satellite image, since there are literally millions of pixels to be transformed.

In addition to being a typical gridding problem, registration is also a transformation from a warped coordinate system to a regular one. A number of common features must be identified on the image and a topographic map that uses the desired map projection. After entering both the pixel and map coordinates for these common points, a transformation matrix can be computed. The transformation is typically a third-

or fourth-degree polynomial, to account for variable distortion across the image. At least thirty control points are needed to calculate a stable transformation matrix. For larger images (e.g., full LANDSAT scenes), more than one hundred might be used. Applying the transformation places all pixels at the correct location in the map coordinate system. Interpolation onto a regular grid in the map coordinates follows; this step is essentially the same as the gridding required in geometric correction.

A final step that is sometimes needed is to create a mosaic from more than one image. Since satellite scans follow set orbits, the ground area covered by an image is fixed. (This is not completely true for newer systems such as SPOT, which can be programmed to scan off the orbit track; however, for standard coverage that can be obtained by anyone after the fact it is a valid statement.) The area to be studied might not be contained within a single image, in which case a composite image must be constructed. This requires careful matching of coordinates, and possibly adjusting pixel values to compensate for differences in the data. The images to be combined may have been recorded at different times, even years apart. (Each scene is selected on its owns merits, considering best contrast, freedom from cloud cover, and other factors.) The balancing of data from each source is based on statistical comparisons of pixel values over similar features from each one. Like many other operations on images, it is at least partly a trial and error procedure. The best measure of success is a seamless appearance in the mosaic; that is, the boundary between the original scenes cannot be detected.

DATA ENHANCEMENT

Although useful information can be extracted from the raw data (by plotting each channel as a gray-scale map), further processing is required for the most effective use. Remote-sensing data are multivariate in nature; that is, the values of several independent variables are known at each location. Here the variables are the intensities of radiation in each of the frequency bands to which the system responds. As noted, LANDSAT records seven bands, and some airborne systems record twelve or more. In the future, this number undoubtedly will increase (systems with more than one hundred frequency bands have been proposed).

The main problem, then, is to find ways to combine the separate bands into a single display, to bring out features not detectable in any single band, and to keep the volume of output at a manageable level. It is in effect the same problem faced in multielement geochemical surveys, except that the volume of data is orders of magnitude higher. Not surprisingly, many of the same statistical techniques are used (as noted in the next section).

There are also many special procedures for interactive display of the results. Even after combining variables, the large number of pixels remain. The only practical way to examine the data is graphically, using a high-resolution color screen and software that allows display parameters to be changed repeatedly. There are two different ways of viewing the data. The first is to take a single variable (either an original spectral band or some derived quantity) and assign a range of colors. This is the same procedure described in earlier chapters for single variable maps. The color assignment may be continuously across the range of the data, or within fixed ranges.

Ideally the software should allow the color assignment to be adjusted with complete freedom, by using a pointing device on a color palette shown on the screen, for example. To make the new display of the image happen quickly, special hardware is required: The color change must be done mostly in the video display, with a minimum of processing. The density slicing technique produces multiple views of the image using narrow ranges. It enhances features with a strong response at the particular energy level.

The second procedure is to use three variables to control the three primary colors of the display unit. This approach allows a combined look at more than one variable, without necessarily having to do fancy statistical operations. The standard false-color images for LANDSAT are produced in this way, using the same bands and color assignments for any scene, so that the visual characteristics are the same.

For either method, additional scaling of the data is often needed to use the full color range available. This process is called contrast stretching, which in essence aims to produce a distribution of values that spans the full range allowed. Several methods are used; for example, normalizing histograms, or converting to a Gaussian distribution. Various type of color transformations are also used to improve the visual effect of the image. Filtering techniques may also be used to emphasize trends in certain directions, or to enhance features within a particular range of sizes.

It should be clear that a program for simply displaying remote sensing or other image data must have a lengthy list of options available to the user. It is essential that different formats can be easily and quickly tested and compared: As in many other areas, no single procedure is guaranteed to work on all sets of data.

DATA ANALYSIS

Multivariate statistical techniques are used to simplify the data and to aid in its interpretation. The initial objective is the same as for multielement geochemistry. Since each frequency band may be considered an independent variable, it takes several displays of selected bands to see all the data (even when using three bands at a time). Because the bands are not fully independent (due to overlapping spectral signatures from the Earth), it should be possible to transform the original variables into a smaller set. Many of the multivariate techniques for reducing dimensionality of the data are applied, with principal components analysis probably being the most popular. Essentially all of the variability in the data might be expressed in three principal components, so that a single color image could be constructed to show all the features of the image. Discriminant analysis and other similar methods also have been used in remote processing, with the same objective.

As in other areas where principal components (or related methods) are used, it is desirable, but not always possible, to establish a physical explanation for the components. In other words, the analysis will indicate that some combination of the original variables is a significant influence on the data, but does not explain the underlying cause of this relationship.

If the remote sensing data are used simply for anomaly detection, this may not be a major drawback. In this case, the objective is to locate areas in the image that do not correspond to known features. The first few principal components generally represent most of the variability in the data, and are likely related to regional features. Local anomalies tend to show up in the higher order components. The search for anomalies is then a matter of selectively displaying each principal component, and looking for features not visible in the basic image. This is similar to multivariate methods in geochemistry, where map displays are used to highlight samples with unusual element distributions. The unusual samples may be selected on the principal component values, residuals from factor analysis, or other similar criteria. As in such other types of anomaly finding, the next step is to examine the anomalous areas of the image with some other exploration method.

Another application of multivariate analysis is classification. The objective is to find areas of the image that have common properties. Some measure of similarity is computed for small subareas of the image, and each subarea is ranked according to this value. For example, considering each pixel as a point in N-dimensional space, the distance of each pixel from the origin may be used as the measure of similarity. Classification attempts to define a set of parallelepipeds (i.e., N-dimensional boxes) containing groups of pixels.

The ranking may be done in two ways. Unsupervised classification assigns all pixels to one of an arbitrary number of groups. The program determines the number of groups automatically, based on the distribution of the ranking variable. The problem of determining what the groups represent exists here as well. The solution is to plot the groups in symbolic form on a map, and then see if each group corresponds to some common features on the ground. For example, one would expect areas of water, trees, and exposed rock to fall into separate groups.

In supervised classification, the identification of ground features is done first. Training areas are defined on the image for the items of interest, for example healthy trees versus environmentally damaged areas. The classification program determines the statistical character-istics of each area, and then examines the rest of the image, to find additional regions that are similar to the training areas. This method is obviously of more immediate use in interpretation. It is not without problems, however. Choice of training areas is critical: The program might determine that no other areas match the predefined groups! In many parts of the world, data with which to compare the results (ground truth) are not available. When tracking time-dependent phenomena, even existing geologic and land-cover maps may not be representative of what is seen in the image.

Another common type of analysis is edge detection (or lineament analysis). The objective is to find linear features in the image, which might be signs of geologic boundaries, traces of buried faults, or other features. This problem is essentially one of filtering, in which the desired outputs are features that are short in one dimension, and long in another. Depending on existing knowledge of the area, the filter may or may not be biased toward looking for linear features in a particular direction. Edge detection often requires an iterative approach with different criteria on what comprises a linear feature. First attempts might produce a vast number of features in apparently random directions.

These types of analysis may be used primarily as aids to mapping. The objective is to create a display that provides the best visual distinction of geological features. In other words, the analysis is part of data enhancement. Where time-dependent phenomena are involved (snow cover, vegetation, etc.), a series of maps showing variable boundaries may be the prime reason for using remote sensing.

Particular applications may use special types of analysis. In mapping water resources, it is possible to estimate ground moisture, snow depth, and similar parameters from the signal strength of spectral bands or radar reflections. Distinctive shapes of spectral curves can distinguish healthy plants from those weakened by heavy metals or acid rain.

INTEGRATION OF IMAGES AND OTHER DATA

Since images from remote sensing are only one of many exploration tools, it is necessary to compare them to other types of data. This requires the ability to display all types of data at the same scale and in the same coordinate system.

The first prerequisite is to geocode the image data; that is, to transform the coordinate system to a standard map projection. As noted above, this registration process is not routinely done by the distribution agencies. Once it has been done, there are two approaches to the problem of comparing the data. First, and simplest, is to produce maps and image displays independently, with a common origin, grid reference, and so on.

This allows two plots to be overlain on a light table, or viewed side-by-side. Since large color displays are expensive, it may be more practical to show the images in a gray-scale format. This approach allows the image analysis program and general mapping program to be completely self-contained.

The alternate method is more difficult in programming, but provides considerably more flexibility. Other types of data are treated as additional variables to be assigned to each pixel. The full range of image manipulation techniques is then available, including correlations and multivariate classification. It will require the other data to be in a gridded format, however. For data that are vector in nature, some image display programs have the option to store vectors, to be overplotted as lines on the image. For example, topographic contours, drainage patterns, or other physical features might be displayed to provide a geographic reference. Similarly, options to store and plot discrete reference points are useful. Typically the display options for these other types of data are somewhat limited, as compared to a conventional mapping program.

To provide a single map with image data and a complex display of other data, a composite output can be created. That is, the image program and mapping program are run independently as before, but their outputs are merged before plotting. Some planning in color assignments, line thicknesses, and so forth is required, so that both types of data will be clearly visible. A raster plotting device (such as an electrostatic or laser plotter) is also necessary. The composite may be done only for hard copy, or in displaying the image interactively. The advantage of this approach is that it avoids extensive duplication of effort in writing the software. The image system needs only to be able to load a graphics overlay, and does not have to provide all of the data manipulation options of the general-purpose mapping system. An example of a combined display is shown in Figure 5.1.

IMAGE ANALYSIS OF CONVENTIONAL DATA

The use of image analysis for other types of data is not restricted to comparisons with remote sensing. The power of image-analysis methods

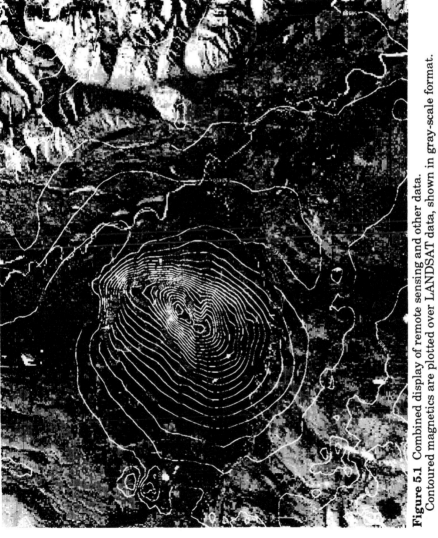

Figure 5.1 Combined display of remote sensing and other data. Contoured magnetics are plotted over LANDSAT data, shown in gray-scale format.

to enhance subtle features, and to manipulate large data sets, is often a benefit in itself. Particularly in geophysical surveys, the efficiency of modern data-acquisition methods allows large volumes of data to be recorded. Conventional programs for data analysis and display may be strained to their limits when the number of data reaches into the tens and hundreds of thousands, which is the norm for airborne surveys, and can be approached with ground surveys using automatic recording.

As noted above, an image is simply a two-dimensional array, so any gridded data can be treated as an image. Some scaling of data values is almost always needed, since image-analysis programs usually require data to be expressed as integers in the range 0 to 255. (This range allows more compact data storage, and computations using integer arithmetic, which is must faster than floating point operations.)

Since many of the applications involve a single variable, sophisticated multivariate techniques are not as commonly used for these types of data. For example, regional structural mapping in mineral and petroleum exploration may employ an airborne magnetic survey as the first stage. Techniques such as filtering and edge detection are used to search for hidden trends in the data. Pattern recognition (classification) and lineament analysis may be useful in comparing the data to geological maps. The ability to try various color displays may be the major benefit.

One method that is useful for enhancing directional trends in the data is sun-angle shading. It exploits the well-known fact that aerial photographs taken at different times of day show different structural features, because of lighting from different directions. This effect can be simulated by treating the data variable as though it were an elevation, and scaling each value by the difference in light intensity that would be expected given the slope of the surface. Stated differently, the intensity on slopes facing away from the light source is reduced, relative to those facing the source. For nontopographic data, some scaling may be required, to provide comparable units of the data variable and the coordinates. If the apparent slopes are too large or too small, the shading may produce little effect (or transform most of the image into shadow).

This technique is applied several times with different sun angles, to selectively enhance features perpendicular to the current direction (Fig. 5.2). The algorithm does not require extensive computations, and

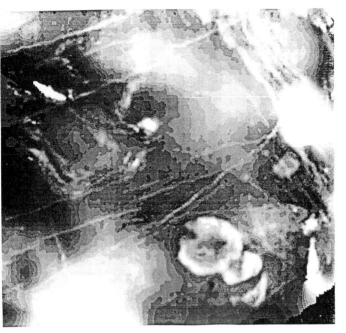

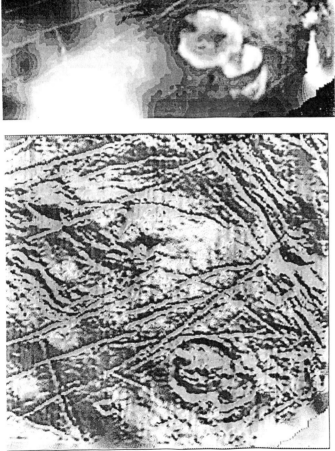

Figure 5.2 Sun shading of aeromagnetic data.
A: Original aeromagnetic data in gray-scale format.
B: Same data with synthetic sun shading. Note greater visibility of many linear features.
(Reprinted from *Computers & Geosciences*, vol. 14, no. 5, 1988, pp. 664–665, "An IBM-compatible microcomputer workstation for modeling and imaging potential field data," by John Broome.)

is inherently stable. It is becoming more popular than directional filtering, which has the same basic objective. With the right graphics hardware, it is possible to do this kind of shading with continuous variations in real time, so that all directions can be tested in a matter of seconds.

CHAPTER 6

Analysis of
Drillhole Data

The indirect methods of exploration are limited in determining what is happening below the surface. Inevitably the need arises to take samples from the target region, to confirm that the interpretation of surface measurements is correct, and to provide more detailed information. In addition, many properties of the Earth cannot be measured indirectly, and remain essentially unknown until a physical sample is available. The primary method for acquiring such samples is drilling from the surface, and recovering some of the material in the drillholes.

Drilling does not provide an unambiguous picture, of course. It is an expensive procedure, and thus only a limited number of holes will be drilled. When considering the volume of material contained in the target region, the amount of sample recovered by even an extensive drilling program is negligible (one part in millions at best). A great deal of interpretive judgment is still required to extrapolate the information gained from the drill samples to the complete zone. This chapter examines the computer methods used to aid interpretations of drillholes. In Chapter 7, the discussion will move on to the most difficult problem of all: estimating the amount of the target resource that is actually contained in the exploration zone.

Notes on terminology:

Resource is not necessarily a positive concept: The aim of drilling may be to define the extent of a contaminated groundwater zone, for example.

The term *drillhole* is equivalent to *well* when discussing oil and water in the ground. In those applications, wells are also used in extraction of the resource, so it is standard to speak of exploration wells and development wells as distinct categories.

In exploration, the aim is to determine whether a resource worthy of development exists, so the recovery of material is a secondary objective. Similarly, in geothermal exploration, the first step is to determine whether the temperatures at depth are sufficient. Later development also uses wells to extract heat energy (typically by pumping water into some wells and out of others after heating).

DATA ACQUIRED IN DRILLHOLES

In one sense, a drillhole is no different from a survey line along the surface. It is just another set of observation points, at which all of the same types of measurements might be taken. Since the drillhole still provides direct sampling of a very limited region (the same as taking soil samples along a surface traverse, for example), many indirect techniques are applied by lowering instruments into the drillhole.

Drill Logs

Depending on the purpose of drilling, geophysical measurements in the drillhole may be the primary type of information acquired. In petroleum exploration, for example, several types of logs (records of geophysical responses down the hole) are used to detect the presence of oil, and to estimate reservoir properties of the rocks. The logs (also called well logs) are recorded automatically, usually by a special computer on the surface that receives signals from the detectors in the "tool" moving through the hole. The logs may record active or passive methods (for example, sonic impulses to determine characteristic seismic velocities, or gravity measurements to detect density variations).

In other cases, the primary reason for drilling is to retrieve rock samples from depth. In mineral exploration, for example, the amounts of economic minerals are measured directly in the rock chips or drill cores recovered in drilling. Downhole geophysical surveys may also be used to try to detect ore zones that may be near the hole, but which were not penetrated by the drill.

In mineral exploration, drill logs are also recorded. In this situation, the normal meaning of log is a geological description of the rocks along the path of the drillhole. The log is developed by a geologist looking at the samples recovered at the surface, and thus does not involve running specialized instruments down the hole. Although the geologic log may be simply hand-written, it is becoming common to use computerized coding formats or lap-top computers to put the geological description into a form amenable to computer analysis. If geophysical techniques are also used, the recorded response is typically described as a geophysical log (to distinguish it from the term *drill log*, which would refer to the geological type).

Drillhole Location

Spatial coordinates are required for all drilling applications. As for all types of exploration data, knowing where the information was taken is crucial to interpretation. For drillholes, location is three-dimensional, unlike some surface work, in which the elevation along survey lines may be ignored. Determining true position in a drillhole is considerably more difficult than for surface sampling, since conventional surveying methods requiring line-of-sight measurements are not applicable. In addition, it is impossible to precisely control the path of the drill, since it will tend to follow zones of weakness in the rocks.

Downhole surveys require lowering special instruments into the drillhole. Unlike geophysical logs, it is impractical to take continuous readings, due to the need for precise positioning and orientation of the surveying instruments. As a result, a relatively limited number of survey control points are taken. At each point, the current direction of the hole is measured. Data for each point consist of depth down the hole, and two angles: the azimuth, or horizontal angle (measured clockwise from north), and the dip (vertical angle). Surface coordinates for the top of the hole (called various names such as collar or kelly, depending on the jargon of the industry doing the drilling) are acquired using conventional survey techniques.

To calculate positions of intermediate points requires some assumptions. The simplest (and most commonly used) is that the track of the drillhole varies smoothly between the survey points. Pseudosurvey points may be generated in between the real ones to allow the azimuth and dip to change gradually, rather than simply interpolating only between the original points. The end result is that the trace of the drillhole is represented by a series of straight-line segments. Since the survey points are measures of local direction, they occur in the middle of appropriate segments (Fig. 6.1). This example has only one computed inflection point between each pair of survey points: Extra values could be added for a smoother curve.

It is a simple problem in trigonometry to determine X-Y-Z coordinates by interpolation between the survey points (whether original or implied inflection points). It is essential to plot the computed position of each drillhole to verify that the survey information is correct. Simple errors such as a sign difference in an azimuth are hard to spot when looking at a list of numbers, but are immediately obvious when a location map shows a drillhole going in the wrong direction. Two views of the hole are needed to properly check all three spatial coordinates; a horizontal plan and a vertical section. An example survey check plot is shown in Figure 6.2. A software package for drillhole analysis and display should be able to plot plan maps and vertical sections with equal ease.

The interpolation scheme outlined above can be incorporated into the package, so that the drillhole data base contains basic survey data only. This allows a more compact file than computing (and storing) X-Y-Z coordinates for every point. A drillhole may have only four or five survey control points, but hundreds of data points. If position errors show up in a plot, it is obviously easier to correct the few survey points than a large set of computed coordinates. Although the coordinates may have to be recalculated many times, this is not a large computing task.

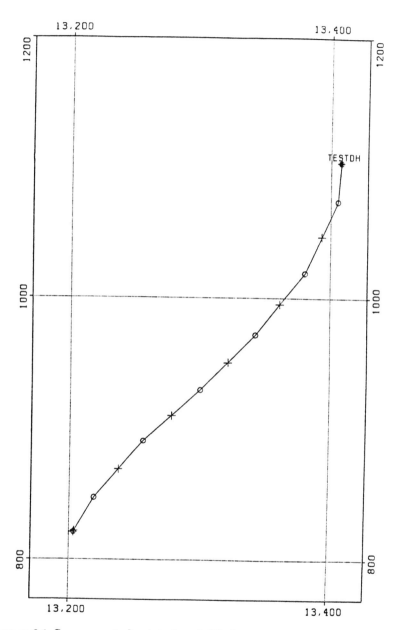

Figure 6.1 Survey control points in a drillhole.
Actual survey points are marked with a +, with the inferred position of
direction changes circled.

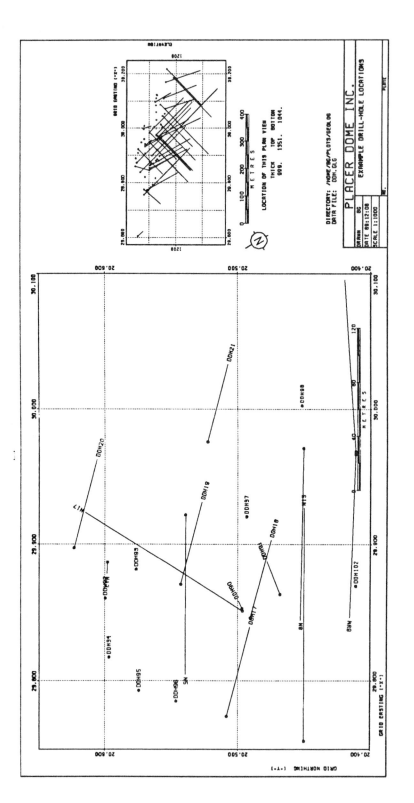

Figure 6.2 Check plot for drill-hole locations.
A surface projection and vertical view are combined so that the inferred position of the holes can be seen in three dimensions.

GRAPHICAL METHODS FOR DATA ANALYSIS

Graphical methods are important in all types of data analysis, but perhaps exceptionally so when working with drillhole data. Because of the problems of maintaining constant drill direction, it is unusual to have perfectly regular drilling patterns. Logistical and economic considerations also may dictate erratic coverage. Offshore drilling often has several holes drilled from the same platform. In late-stage mineral exploration, an underground tunnel may be used to provide drill sites close to an ore body, with drillholes fanning out from settings within the tunnel. It is important for data displays to show the drillholes in their correct spatial relationship, so that clustering and gaps in sampling can be accounted for in interpretation.

Computer-Generated Cross-Sections

The primary way of looking at drillhole data is the vertical section (often just called a section, or a cross-section if it cuts across the smallest dimension of a structure, or a longitudinal section if it is drawn along the long dimension of a structure). A section is a two-dimensional view, where the vertical axis is elevation (or depth below the surface or some standard reference elevation), and the horizontal axis is distance along the line where the plane of the section cuts the surface. In some cases, the vertical axis of the plot represents distance along a dipping (i.e., nonvertical) plane. This variation is useful when the drilling target is a tabular body that is not vertical or horizontal in the crust (a fault zone, for example).

In the most general case, we might consider a section to also include horizontal views, which would more traditionally be called maps or level plans. These might include data only within a specified elevation (or depth) range, or might show all data projected onto a horizontal plane. The same criteria may be applied to other orientations of sections as well; for example, a vertical section typically shows all data within a set distance of the projection plane.

To construct a section (manually or using a computer) is largely a matter of doing trigonometric calculations to project each drillhole onto the plane of the section. This is a special case of translation and rotation of coordinate systems. The normal coordinates for each data point in the drillhole are converted to a system aligned with the section line. Vertical sections require a two-dimensional transformation, with the elevation coordinate being unchanged. For nonvertical sections, a full three-dimensional transformation is needed. Note that when the section line is parallel to either axis of the survey grid, the transformation is just a simple matter of choosing one of the map coordinates for the in-line distance (perhaps with a sign change to allow for the desired viewing direction). For example, an east-west section (looking north) may be drawn using the normal easting and elevation as plot coordinates.

There are a variety of ways to specify the location of section line for use by a section-drawing program. The end points of the section line may be entered by simply typing in their coordinates. In addition, the viewing direction must be defined: Two (visually) different sections may be drawn for the same line, for a hypothetical observation point on either side of the line. Alternately, the starting point of the line may be specified, along with its direction and length (all of these possible parameters are interrelated, so that a line may be uniquely defined by only some of them). Another variation is to request a section line parallel to a given line, at a particular perpendicular offset. This procedure is convenient when drawing a series of sections that are not aligned with one of the coordinates axes, since it saves manually calculating a series of end points for each line.

Instead of manually picking end points from a drillhole location map, a section-drawing program might allow this to be done graphically. The drillhole map is displayed on the screen, and the end points of the line are defined by moving a cursor to the desired position. Depending on the hardware configuration, the positioning may be controlled by a mouse, digitizing tablet, or arrow keys on the keyboard.

When the drilling covers a broad area on the surface, a maximum distance of projection is imposed, so that only holes that are close to the section plane will be displayed. In this case, many parallel sections may be drawn, each taken to represent a thickness equal to the spacing of the sections. (Section thickness has implications in computing volumes and estimating resources, as we shall see in next Chapter 7.) Figure 6.3 is such a section, where a fixed width is used to select the holes to be plotted. This selection is straightforward once the coordinate projection is done,

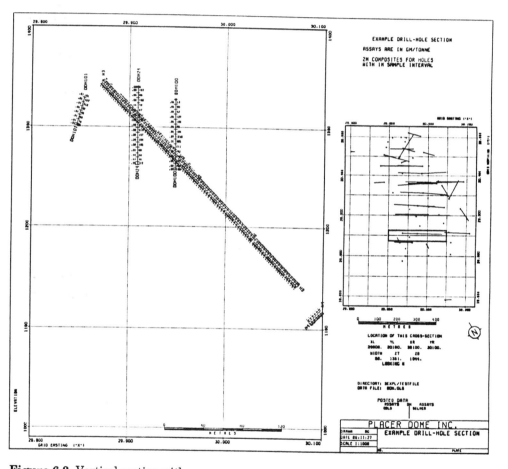

Figure 6.3 Vertical section with
limited width.
The inset map shows the area
included in the section.

because the distance from the section line is one of the transformed coordinates.

After the location of the section is set, the next problem is to determine which data to plot for each drillhole. This is obviously dependent on the types of data available, the preferences of the interpreter, and a variety of other factors. Almost inevitably compromises are required, since it is very difficult (if not impossible) to plot all of the available information at a practical scale without making it unreadable. Many of the symbolic coding schemes and other graphical techniques used in geochemical and geophysical data analysis are applied in sections, as much for the data compression effects as for highlighting anomalies. (At this stage in the exploration process, the objective is more to gain a detailed knowledge of the subsurface region than to find new targets.)

The volume and great variety of data can also pose problems in writing an effective section program. It is essential to find simple, efficient, and flexible ways for the user to define the types of data to be plotted. There is no basic formula, except that continued cooperation between the software developer and the users is mandatory. Good interactive programs for such a complex and specialized task only arise from extensive use by several people, and good communication of problems and suggestions back to the person responsible for programming.

A recurring problem in plotting cross-sections is overposting. It also occurs in many mapping applications, of course, but may be more common in drilling programs due to drilling fans from the same setup point. A general solution to overposting is very difficult, but some relatively simple measures can at least reduce the problem. Often excluding one or two holes can greatly reduce the clutter. A program can easily provide this option by displaying a list of holes to be plotted before actually starting to plot them. The excluded holes might then be plotted separately, or moved to the side of the main section plot. In either case, it is useful to show the location of these holes (just as a line showing the trace of the hole) on the main section, to make it obvious where additional data are available.

Even without considering overlapping holes, the close spacing of samples along each hole may lead to overposting at the desired scale.

Once again, procedures discussed earlier for maps are also effective. In general, use of symbolic or graphical forms lead to more compact plots than posted numbers or text. Many geologists prefer to see original data on sections, though, so there inevitably will be demands for a section program to move text around to fit as much as possible on the plot. Figure 6.4 shows a variety of ways that data can be displayed to reduce overplotting.

The only real answer to this problem is to use interactive plot editing (i.e., a computer-aided design program) to modify the plot created by a fully automatic program. Unfortunately, this method may not always be practical, since it is not unusual to plot many cross-sections for large drillhole patterns (thirty or more sections are commonly needed in mineral deposit drilling, and in some cases the number might easily be in the hundreds).

To solve the overposting problem automatically, it is necessary to consider the entire plot as a large grid with very fine subdivisions, similar to individual pixels in image processing. The plotting software then must keep track of which cells have already been filled whenever new information is added to the plot. Conceptually this is a simple matter of maintaining a counting array, with some searching mechanism for finding the nearest open area to the desired point. In practice, it is difficult, since the desired resolution of the plot requires many cells. The need to maintain the counting array demands large memory, and adds a great deal of overhead to the basic plotting software. Especially on smaller computers, these demands cannot be met without sacrificing too much speed (if they are even within the capability of the machine). As a final note, many sections (and maps) are built up in successive layers with different types of data, so that any such scheme might eventually fail simply because of the density of information displayed.

Profile Plots of Drillholes

Due to the problems outlined above, views of multiple drillholes on a section do not usually show all data associated with each hole. When detailed analysis of each hole is needed, additional plots may be generated showing each hole by itself, so that many types of data may be plotted side-by-side without overlap. This type of display is usually

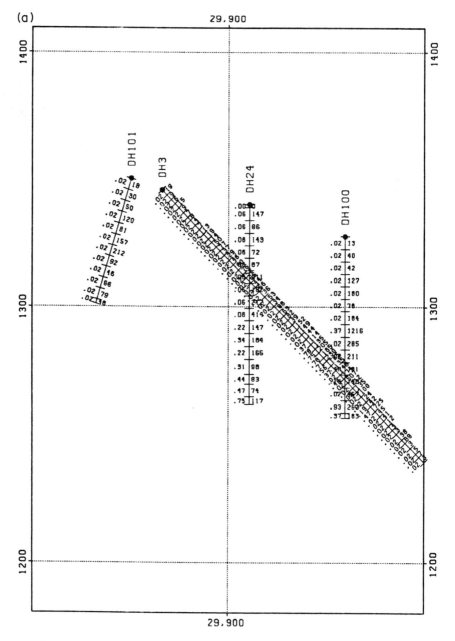

Figure 6.4 Display formats for reducing overposting.
  A: Original section with detailed posting.
  B: Posted form, showing only values above a cutoff.
  C: Pseudohistogram form. Data values represented by the offset from the drillhole trace.
  D: Patterns instead of text.
  E: Exploded views. Portions of the section are plotted twice to avoid overlaps.

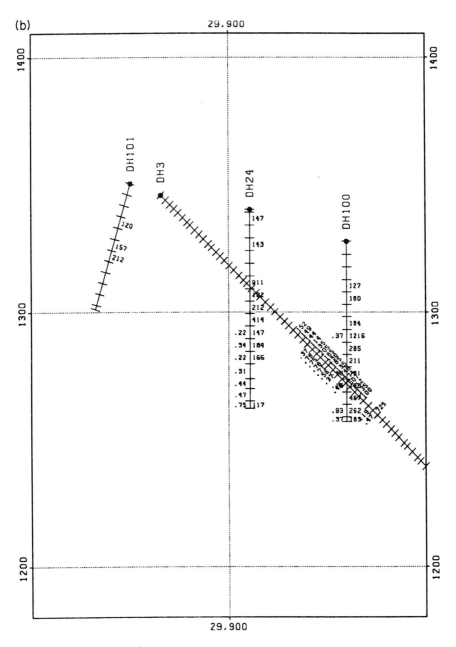

Figure 6.4 Continued.

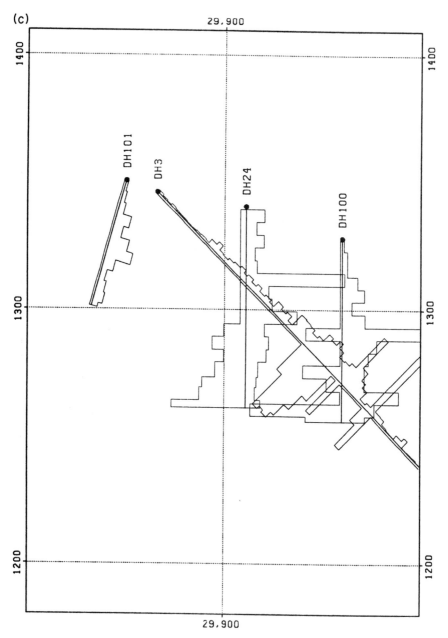

Figure 6.4 Continued.

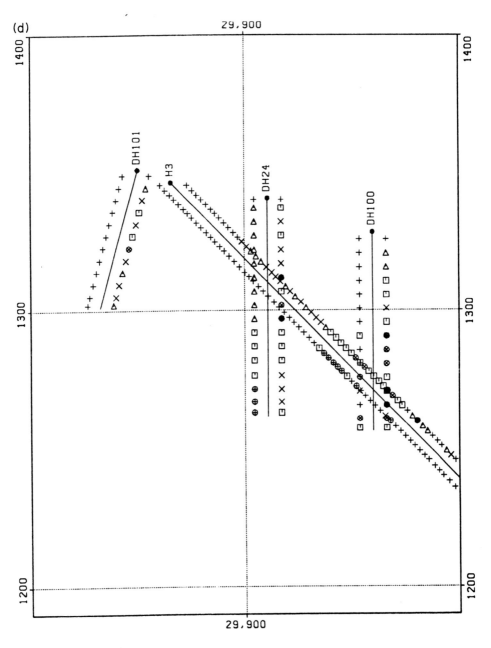

Figure 6.4 Continued.

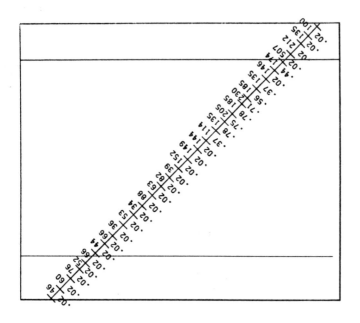

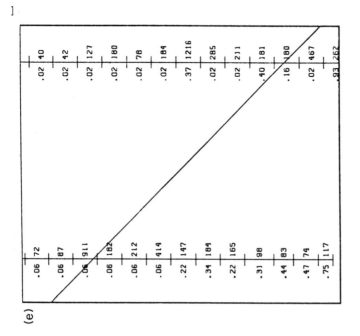

(e)

Figure 6.4 Continued.

created as a series of vertical strips, without regard for the actual orientation of the drillhole (i.e., the plot is treated as a one-dimensional view using in-hole depth as a coordinate, rather than as a two-dimensional view using coordinates projected onto a plane).

The display formats used for graphical drill logs are essentially the same as for other types of plots. Numerical data may be represented by simply posting the numbers, as scaled symbols, or as profiles along the trace of the hole. Qualitative data (type of rocks or minerals, for example) might be shown by text codes or graphical patterns.

Figure 6.5 shows a detailed drillhole plot. Note that the form is similar to that used in the section examples, but that more information is shown. The increased data content results from the smaller scale, and the ability to expand the area taken up by the plot without conflict.

A program for generating such plots may be integrated into a section-drawing package, since many of the functions are identical. For example, projection of numerical data into a profile, or conversion of qualitative data into patterns are fundamentally the same, except for minor differences in the plotting coordinates. To be effective, a profiling program should allow the layout for any hole to be preset, and then simply repeated for a list of drillholes. Use of interactive plotting will help in creating the first plot. Repeating the process for many other holes might be done on the screen, or set up as a background task with output to a plotter or printer.

NUMERICAL PROCEDURES TO AID IN INTERPRETATION

The various graphical displays are not an end in themselves, but are primarily aids to developing a geologic model of the region. Other computer techniques are also useful at this stage.

Correlations Between Variables

One of the key problems is to determine the interrelationships of the different types of data. For example, the responses of various geophysical instruments might be used to determine the type of rocks,

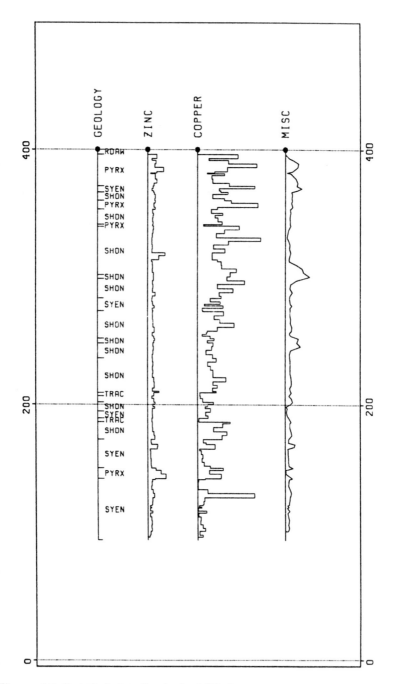

Figure 6.5 Detailed plot of a single drillhole.
Note the trace labeled MISC, which is plotted as a continuous curve connecting point readings in the drillhole. Other traces shown in block form represent average values within the intervals.

or to predict the presence of oil or an economic mineral. Although the relationship may be detected visually, more often it is obscured by other factors (i.e., the geophysical response is not solely controlled by the rock type). Statistical methods are needed to sort out these problems.

One might ask why such methods are necessary, when the examples above showed rock types as known information? In many applications of drilling, geologic logs are taken only for a portion of the holes, and geophysical measurements are used to estimate the geologic parameters in the others (since the geophysical logs are automated, they are obtained faster and at less cost than a geologic log). The assumption then is that relationships found in holes where both geologic and geophysical logs are available also will exist in holes where only the geophysical logs were taken.

In petroleum exploration and development, many sophisticated types of numerical analysis are applied to geophysical data, under the general classification of well-log analysis. The techniques rely on extensive use of graphical displays, as well as correlation and modeling techniques. Since much of the processing is done on individual wells, the analysis techniques are essentially one-dimensional (the single coordinate is depth in the well). The methods of time-series analysis thus are directly applicable.

To see if two different variables are related, the simplest approach is to determine the correlation coefficient between them. This calculation should be available in any basic statistics program, in essence involving calculating a straight-line fit between the two variables, and measuring the average deviation of the data points from the line. This method does not take into account spatial relationships, however. Considering the profile plots of two variables, one may be shifted with respect to the other. For example, oil and gas content may be strongly related, but since gas rises above oil, the association may not be indicated by a simple correlation coefficient. (These quantities usually are not measured directly, but inferred from the different geophysical logs that have distinctive responses to the presence of oil and gas.) Calculating the cross-correlation function compensates for such effects, by testing relative shifts between the profiles to determine the best alignment.

As in most other areas we have discussed, the computer processing of well logs does not produce definitive results; an experienced analyst is required. As before, the key to success of programs for well-log analysis is that they are flexible, easy-to-use, and able to process many logs with the same parameters. An interactive display is essential to allow the analyst to test different methods (e.g., different display formats, combinations of different logs, different assumptions about the geology and reservoir characteristics).

One promising new use of computers in well-log analysis is expert systems (of which more will be said in Chap. 8). Programs of this type aid the analyst in making critical choices of parameters by comparing current information to a knowledge base. The knowledge base is developed initially by expert well-log interpreters, but may be extended by saving results from previous runs of the program. One objective of such developments is that sophisticated analysis may be done on-site, immediately after the data are recorded, without the presence of a highly experienced analyst. The results are reviewed later, of course. The key point is the time savings in getting the information, which allows drilling decisions to be made while the equipment is in place.

Compositing Data

The volume of data in a drillhole may in itself make interpretation difficult. Assays for mineral content may be taken every one or two meters, for example. The sample spacing for geophysical logs (i.e., the interval between instrument readings) is usually even shorter. The result is hundreds or thousands of data points for each variable in each drillhole. The interpreter's aim is to develop a simplified model from these data, so it is necessary to combine the individual data points in some fashion.

There are three methods for compositing data. First is a simple average of adjacent values, which increases the effective sample length of each value. For example, combine five 2-meter samples into a 10-meter composite. In the event that the original samples were not all of the same length, a fixed-length composite is calculated: All data values within a constant length of the hole are combined by a weighted average, to account for the variable length. One original sample might contribute to two composited samples, depending on where the sample and composite intervals are in the hole. Therefore, the sample data values are used in both composites, with a prorated length. A sample from 19 to 22 meters would be treated as a 1 meter sample for the 10- to 20-meter composite, and a 2-meter sample for the 20- to 30-meter composite.

The second method is compositing within specified intervals, which typically represent different geologic units. Computationally this is nearly identical to fixed-length compositing, except that the intervals are of different lengths. The top and bottom of each zone must be entered by the analyst. These might come from a file of zone boundaries picked from the graphical logs of each hole, or may be picked interactively from a display of the holes on a screen. An identifier for the zone may also be attached to the output file, to allow different zones to be separated for statistical analysis and reserve calculations (discussed in the Chap. 7).

The final method is compositing within horizontal or vertical slices of constant width. It is used when the reserve estimation method employs blocks of constant size. In open-pit mining, these are called bench composites, since the ore is extracted from benches of fixed height. The entire orebody is treated a a set of horizontal layers, each corresponding to a mining bench. Similarly, the body may be treated as a set of vertical slices, for which the portion of each hole within each slice is to be averaged. The intersection points of the top and bottom of each slice must be found in the drillhole, prior to applying the normal procedure for compositing within zones. This can be done automatically, of course.

Now the intervals for compositing are not necessarily constant along the drillhole, since the holes may not be perpendicular to the slices. In the event that the drillholes are at a shallow angle, it is customary to revert to fixed-length composites along the hole, to avoid averaging along an excessive length of the drillhole. In the extreme, the entire hole might lie within one slice (e.g., drilling horizontally into a mountain, or using vertical slices with vertical holes). For example, if the composite height is 10 meters, and the hole intersects the zone at 30 degrees, the length of composite along the hole is 20 meters. For angles less than 30 degrees, 10-meter composites along the hole would be used.

For all types of composites, the output is a file with the composited data values, plus coordinates of the midpoint of the interval, plus associated data such as zone name and in-hole interval. This output is used primarily in computing reserves, but also provides another way of reducing clutter on sections. Figure 6.6A shows the same section as

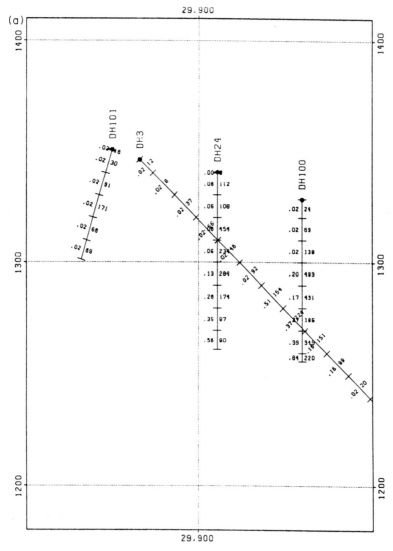

Figure 6.6 Section with composited data.
A: The same section as Figure 6.4A, showing 5-meter composites instead of original samples. Note the improved readability.
B: Combined display of composites and original data. On the left, composites are posted, with original values shown graphically. On the right, the convention is reversed.

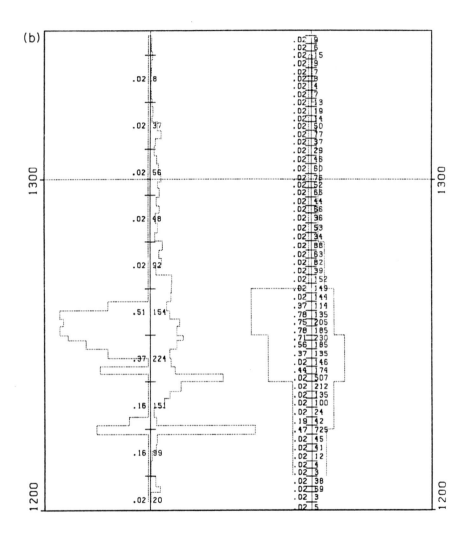

Figure 6.6 Continued.

Figure 6.4, but with composited values. It is useful to show the variation of individual values within the composite, as shown in Figure 6.6*B*.

Depending on the software used for plotting sections, it may be necessary to put the composited values back into the drillhole data base. The program might require the use of original survey points instead of the individual coordinates for each composite. It might be difficult to plot original and composited values together if they are in separate files with different formats.

There are other benefits from integrating the composites, in addition to the ease of plotting sections. All data for a project may be maintained in a single file, which greatly reduces the chances of losing data through inadvertent deletion of apparently redundant files. Procedures for manipulating the composites will be identical to those for the original data, since they are now in the same format. Whenever different formats are used, variations in the user interface are needed to retrieve the desired data from the file. Differences visible to the user can be minimized in the applications program, but at a large expense in programming. Writing a routine to integrate the data is easier. It also makes for more efficient execution of the analysis routines, since options to support a variety of input formats can be eliminated.

If separate packages are used for data analysis/plotting and evaluation/modeling, it may still be necessary to have different forms of output for the composited data. Alternate formats can be provided by the compositing routine, or by a simple stand-alone reformatting function. As noted above, there is little to be gained in modifying the major systems to handle different data structures. When the software is purchased (not developed in the organization), it is usually not practical to modify the internal data structures anyway.

CHAPTER **7**

Resource
Amount Estimation

The final stage in many types of exploration is trying to determine the quantity of the material for which one has been looking. *Quantity* may be as simple as the volume of a particular geologic zone, or as complex as the potential profit to be gained after extracting a variety of minerals from the zone. It is worth noting again that *material* can have a detrimental quality, such as trace-metal contaminants in a supply of groundwater.

There are many uncertainties in the estimation process, and it is difficult to judge which method will give the best results. Indeed, it is often impossible to make accurate comparisons even after the fact, as unforeseen changes in the exploitation method may result in extraction of different zones than originally intended. Sometimes the extraction method itself involves large-scale blending of material, so that the amounts from particular zones cannot be determined.

In gold mining, for example, much of the lower grade ore is stacked in huge piles, and the gold is extracted from the entire pile by leaching with appropriate chemicals. It is possible that material from several levels in a mine will be combined, and only the total amount of recovered gold will be known. As a result, estimates of gold content in individual mining blocks can never be confirmed.

In petroleum and mineral exploration, this process is called reserve estimation. This term comes from considering material remaining in the ground to be a reserve of material available for future production. Reserve estimation is an ongoing need in an operating mine or oil-field, since information gained at various stages of production can be used to refine the estimate of what is left. Changing economic circumstances also can demand revisions to a reserve estimate, since by definition a reserve is the amount of a resource that can be extracted at a profit.

Although reserve estimation may start with a calculation of the volume of the zone, in general the problem is more complex. The volume is certainly important in defining the total amount of the resource, but many other factors are even more important in determining the economic viability of extraction. In oil fields, for example, the flow characteristics of the oil in the rocks may vary greatly within the same reservoir. A detailed model of how the oil will move in response to pumping, water injection, and other techniques must be developed before a reliable estimate of the amount of oil that can be recovered can be made. (Note that this estimate is different from the amount of oil in place in the ground.) The variation of porosity and permeability throughout the reservoir must be estimated in developing such a model.

Similarly, a mining ore reserve requires a detailed model of how ore grades vary throughout the deposit, since the economics are dependent not only on how much metal will be produced, but also on when it will be produced. The objective in planning a mine (or any other capital-intensive project) is first to pay off the start-up costs, which means it is best to have the highest rates of production in the early stages.

As this brief introduction indicates, the estimation phase of exploration requires sophisticated computer applications. It is another area requiring specialized skills and experience, and thus is somewhat beyond the realm of generalized software that can be used in a variety of situations. Detailed resource estimation is generally done by specialists, although preliminary studies are often done by exploration groups as an aid in planning further work on a project. An understanding of the procedures and problems of reserve estimation helps the exploration staff provide the necessary data to the specialists. This chapter will be largely an overview of the types of procedures involved in this area, emphasizing how computerized estimation may be used as an extension of the techniques discussed in previous chapters.

ORE RESERVES IN MINERAL EXPLORATION

An ore-reserve estimate has two distinct components: the volume of the ore-bearing zones, and the distribution of the economic minerals within them. The first relies heavily on a geologist's interpretation; the actual calculation is relatively simple after the limits of the ore zones have been defined. The second can involve more complicated computations, and careful checking that the geological interpretation is also properly applied. Both procedures generally must be based on limited data, and so a high level of uncertainty in the results is almost inevitable.

Because of these uncertainties, and the fact that the reserve estimates cannot be fully validated, there is no correct method for doing estimations. Even in an organization that may have a strong preference for one particular method, it is common to do more than one estimate using different methods to increase the level of confidence in the result. This in itself is no guarantee of reliability, however. Two different computational methods may produce equivalent results, both of which may be far off the mark, due to a faulty geologic model. The inherent uncertainty of the problem in itself may make it difficult to choose one estimate over another.

A key factor in effectively using a computer for estimation is that the software must allow a variety of methods to be run easily on the same data. In addition, the reserve program must be used in conjunction with general-purpose statistical and plotting software to allow easy comparison of the different results.

Before proceeding to a discussion of techniques, a couple of brief notes on terminology are needed. *Grade* and *assay* will be used interchangeably to mean the measured concentration of a metal (or economic mineral). Where the discussion refers to a single assay, the possibility of repeating the same procedure for other metals (or for derived measures such as ratios) should be taken for granted.

Traditional Sectional Method

Like many other aspects of mineral exploration, the first level of computer application is simply to automate manual techniques. In calculating reserves, the manual approach requires the geologist to divide the ore deposit into a relatively small number of zones, each with a well-defined boundary. A single zone might be the outline of the ore deposit as seen on a single cross-section, for example. The complete deposit is defined by a series of cross-sections, as illustrated in Figure 7.1A. Defining these zones is the essence of the geologist's job in interpreting the drillhole information. Each ore zone is drawn over the drillhole display, as shown in Figure 7.1B.

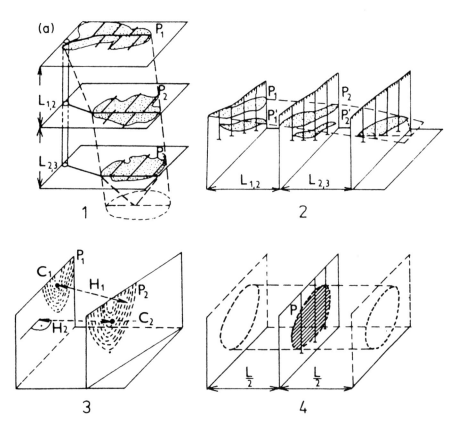

Figure 7.1 Ore zones on sections.
A: Basic principle of the sectional reserve method. (Reprinted from Prospecting and Exploration of Mineral Deposits, by M. Kuzvart and M. Bohmer, Elsevier, 1986, p. 458.)
B: A simplified ore zone drawn on a drillhole plot.

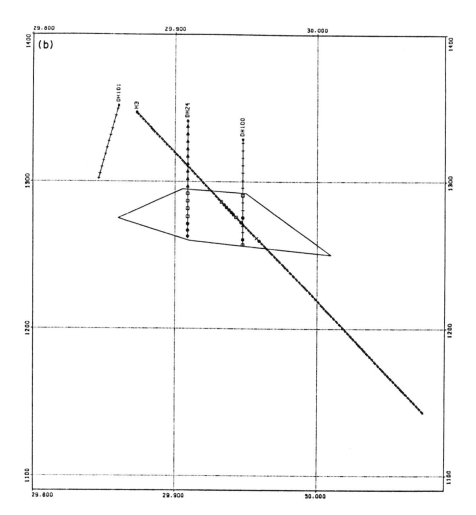

Figure 7.1 Continued.

The first stage is to determine the volume and mass of the ore deposit, for which the area and thickness of each zone must be determined. The area is measured by considering the zone boundary as a polygon, whose vertices are marked with a planimeter (the manual method), or are entered into a computer file using a digitizer. The thickness may be a simple constant in the case where the deposit has been divided into a series of two-dimensional slices. The geologist may have drawn the ore zone on a series of cross-sections spaced 25 meters apart across the deposit. Each section would then be taken to represent a thickness of 25 meters. The deposit may be viewed as a series of horizontal slices (level plans), which similarly would represent a thickness equal to the vertical spacing. If the rock density (i.e., specific gravity) varies significantly within the deposit, each zone will have a specific gravity assigned, otherwise it is assumed constant for the entire deposit.

The next step is to calculate the average of the ore grades within the zone (by one of a variety of methods to be discussed later). Thickness also may be a variable, in the case of tabular orebodies such as veins or coal seams. If so, the average thickness of each zone must be determined in the same fashion as the average grades.

Volume and weight of rock contained in each zone are calculated by simple arithmetic (area X thickness X specific gravity). This is most commonly reported in tonnes (1,000 kg) or short tons (2,000 lb). The amount of ore in the reserve then is obtained by simply adding the tonnes of each zone. The other critical parameter of the reserve is the average grade, calculated by weighting the grades of each zone by the tonnes of contained rock (to ensure that larger zones have an appropriately larger influence on the average grade).

These procedures are easy to automate, provided that the drilling results are already in computer files. Digitizing the ore zones is the most time-consuming job (at least from the human standpoint). The area calculation may be done immediately, or done as a postprocessing job after all of the zone boundaries have been digitized.

In addition to computing areas, the polygons also may be used in extracting the assays within the zone from the drilling data base. In vertical sections, the coordinates of the polygons are the distance from the section origin and the elevation. A section reference number is also assigned, so that the third spatial coordinate can be computed. For example, an east-west section at northing 2,000 is given a label 2,000. The in-line coordinate is the grid easting. When the section is not aligned with the coordinate grid, a cross-reference is defined between the section label and its location on the grid. This enables a standard coordinate rotation and translation to assign grid coordinates to the polygon. Selection from the data base is then a combination of points within a two-dimensional polygon, and a simple minimum-maximum test on distance from the section line.

After extracting the assays for each zone, a single average value is calculated. It might be a simple average of all assays. If the drillcore samples are of different lengths, a weighted average based on sample length is preferable. If the zone is large and the grade distribution is erratic, more sophisticated statistical techniques may be needed, for example, geostatistical methods applied along the drillholes.

Alternately, the required average may already have been determined by one of the compositing procedures described in Chapter 6. The zones will be numbered in both the composite file and the digitized boundaries. The tonnes and grade are then joined by matching the zone numbers.

It should be obvious that a program can combine all the necessary steps. Given a set of digitized boundaries, and a file with assays and coordinates, a list of zones with volumes, tonnes, and average grades can be created without further input from the user. The results are in a simple column format, which may be loaded into a spreadsheet program for doing sums, applying cutoffs to define higher grade zones, printing reports, and so on. (Spreadsheets will be discussed in somewhat more detail in Chap. 8.) Alternately, these reporting functions might be integrated into the reserve program.

Polygonal Method

If the data density is sufficient, it becomes possible to infer the boundaries of the ore zone from the data, rather than from an interpreted outline. This allows a more automated approach to calculating ore reserves, since the time-consuming process of drawing ore-zone boundaries and digitizing them is eliminated. The geologist will have to confirm that the inferred boundaries make geologic sense, however.

The applicability of such an approach also depends on the nature of the ore deposit. It is practical when the ore is not confined to particular structures or formations. For example, many of the large copper mines in western North America have the copper ore fairly uniformly spread throughout large porphyry bodies. The limits of the ore zone are determined primarily by the copper grades, which means that the boundary can be estimated by finding where the grade falls below some particular cutoff value. The cutoff is determined by a variety of factors, but in essence is the grade required to make it profitable to mine and process the ore. It can be different for each deposit, since it depends on many different costs of building and operating the mine. It also has a time dependence, as falling metal prices can change valuable ore into uneconomic waste rock. (Fortunately for the mining industry, this can also happen in reverse.)

When there are structural or stratigraphic controls on the ore zones, a totally automatic definition of boundaries is impossible. In such cases (probably the great majority of ore deposits), a combined procedure can be used. The automatic estimation procedure is constrained by some predefined boundaries. A barren dike may cut through an otherwise continuous body. By digitizing the boundaries of the dike, it can be excluded from the estimation.

The simplest method for automatic definition of ore zones is the polygonal method. Like the section method, it is applied to a series of slices through the deposit. Usually these are horizontal level plans, but they can also represent vertical or inclined sections. For each slice, the data from each drillhole are composited to give a single sample point. The area of influence of each point is determined by drawing a polygon around it. The polygons are defined by finding the midway points to all the adjacent points (Fig. 7.2). The calculation of volume and tonnes proceeds as for sections; compute the area of the polygon, multiply by the thickness and specific gravity, and add the values from each point. Note as before the zone thickness and specific gravity are typically constants, but can be variables that are defined in the drillhole data base. Since the assays have already been composited, the grade to assign to each polygon is known, and the average grade is derived by weighting each one by the tonnes in its zone of influence.

Since almost all ore deposits have internal zones that are not ore, some selection must be done in tabulating the results. This process is also handled most easily by loading the results of the polygon generation into a spreadsheet, as discussed above for the sectional method. However, not all zones will be considered ore, so cutoffs based on the assays are used to select the polygons that are considered ore. This value might be a simple minimum based on the grade of one metal, or might

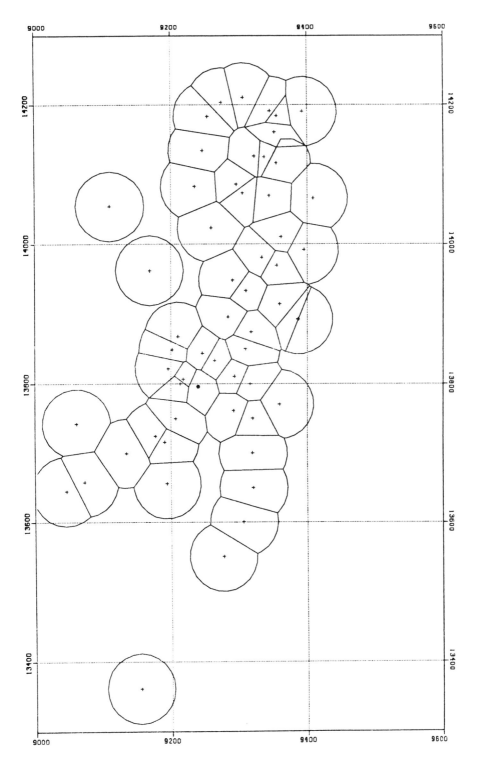

Figure 7.2 Polygonal areas of influence.

be a fairly complicated calculation involving several metals, and perhaps other selection criteria such as maximum levels of some undesirable contaminant. This type of selection is applied in almost all other types of ore reserve estimation as well.

There are two major weaknesses in the polygon method. First is the assumption that the drilling information is sufficient to completely define the orebody. This is seldom the case in practice, so it is necessary to put some restrictions on the generation of polygons, to keep from extending the area of influence of a single drillhole to an unreasonable distance. A maximum distance test can be added to the polygon generation algorithm. If no other assays are found within that distance, the polygon boundary is replaced by a circular arc, as shown in Figure 7.2. This simple solution is not fool-proof, of course. The maximum distance will be used on all drillholes on the periphery of the drilling area. If many of the assays for these holes are above the cutoff, their tonnes are dependent mostly on the choice of maximum radius; and in the total tonnes may be dominated by contributions from these partially uncontrolled samples.

The choice of maximum distance is also based on the distance between drillholes in the middle of the pattern. Here the aim is to ensure that no undefined areas are left within the zone; that is, the maximum distance should be greater than half the greatest spacing between drillholes. This distance is frequently much greater than the maximum extrapolation criteria applied to the peripheral holes, suggesting that two different tests should be used; one for the internal case, and one for the boundary. The polygon generation algorithm should be able to distinguish between these two cases by scanning the directions where the other drillholes are found. This is similar to the octant search procedure used in gridding programs.

Another way to safeguard against unreasonable extrapolation is to use a digitized outer boundary. After generating polygons as normal, their coordinates are compared to the digitized zone boundary. The program must test for an intersection of two arbitrary polygons, and adjust the boundaries of the area of influence if an intersection is found. The same procedure may be applied more than once, if there are additional areas to be excluded from the volume calculation (barren dikes, for example).

The other (and even more serious) weakness of the polygon method is that a single assay is taken to represent a large volume of rock. It is well known that ore grades of almost all metals are highly variable, so it is highly unlikely that any individual assay is an accurate measure of the average metal concentration in the larger volume. Because the range of values for individual samples (i.e., small volumes) is greater than for larger blocks, grade estimates from the polygonal method will be higher than the true value. This error is caused by selection of values above a cutoff, which produces a biased subset of the full distribution.

The use of individual assays (or at best composite assays from single drillholes) is fundamental to the polygon method, and there is no corrective procedure that can compensate for it. Various empirical correction factors may be used for individual mines, but they simply account for the observed discrepancy of ore reserve grades with actual production, and cannot be applied to other deposits. For this reason, procedures that estimate averages over various zones were developed, and have proven to be much more reliable. The polygon method is still used, though, because of its conceptual simplicity (which makes it easy to use and easy to check the results), simple computations (and thus greater speed), and longer history (people tend to prefer the methods with which they are most familiar).

Creating a Block Model

To overcome the problem of using assays directly, a statistical estimate of the average grade of the ore deposit is required. It is usually implemented by treating the deposit as a large number of blocks of regular shape (most often cubes or rectangular prisms). The block model consists of a set of two-dimensional grids, each representing one slice through the model (Fig. 7.3). The blocks may be defined to a fixed shape determined from the geological interpretation. The average grade of each block must be interpolated from the assays in the drillholes.

Assigning values to each block is in effect another gridding problem, subject to all of the difficulties of gridding in general. Because of the three-dimensional nature of the grid, the problem is in fact more difficult than most, since data continuity in the third spatial dimension must also be considered. The searching procedure that selects data values must

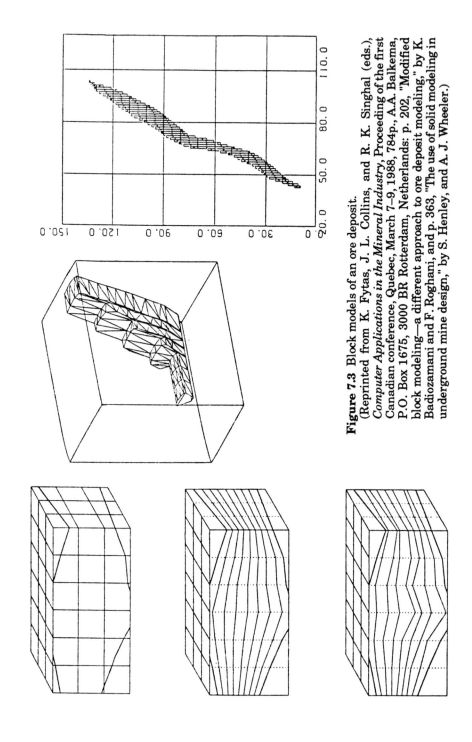

Figure 7.3 Block models of an ore deposit. (Reprinted from K. Fytas, J. L. Collins, and R. K. Singhal (eds.), *Computer Applications in the Mineral Industry*, Proceeding of the first Canadian conference, Quebec, March 7–9, 1988, 784p., A.A. Balkema, P.O. Box 1675, 3000 BR Rotterdam, Netherlands: p. 202, "Modified block modeling—a different approach to ore deposit modeling," by K. Badiozamani and F. Roghani, and p. 363, "The use of solid modeling in underground mine design," by S. Henley, and A. J. Wheeler.)

look above and below the level of the block being estimated, as well as radially around it. (*Above* and *below* should be taken as relative terms, simply meaning the layers of grid cells in opposite directions from the current slice.)

Another problem in estimating grades for all the blocks is simply that there are an enormous number of them. A typical model for a large deposit might have 50 levels, with 100 to 200 blocks along each axis. Thus, there could be between 500,000 and 2,000,000 blocks to be computed. While the computer time may not be a real difficulty, there is no simple procedure for verifying that the estimation process has done a good job of representing the data. Obviously it is impractical to scan all the numerical values. Some statistical comparisons with the original data may be used to guard against gross errors. The fine details of the block model must be checked as well, since it will be used in developing the mining plan. Even producing a set of contour maps of each slice will be time-consuming. A detailed look at the results is very important, since major economic decisions will be based on the block model.

These difficulties have little to do with the method for calculating the block grades, which is another area where many techniques have been used, with varying success. The variety of methods in fact leads to an even larger verification problem, since it is very common to try two (or more) different estimation procedures. This means that two or more full block models are created, and a detailed comparison of the results will be needed. Once again some statistical measures provide a gross comparison, but the choice of the best method will depend on local discrepancies.

As for all gridding problems, after selecting the data for estimating each block, the block value is derived as a linear combination of data values. Compared to many gridding applications, the method of assigning weights is somewhat different, since the value is to represent the average data value within the block, rather than the best estimate of the data value at a point (say, the center of the block). In inverse distance and related methods, the weighting function is adjusted so that all data are considered, even when a single value occurs right at the block center. The averaging approach also helps compensate for the tendency of metal grades to be highly variable within small volumes.

Calculating the ore reserves is straightforward, once the block model has been created and verified. Like the polygonal method, selection criteria must be applied to separate ore from waste, or perhaps to define multiple categories of ore. These groups may also be shown on various graphical displays, to aid in the verification process (Fig. 7.4). Tabulation and reporting of reserves might be done with a special program (as part of block selection, perhaps). For large models, it may not be possible to use a standard spreadsheet program, due to size limitations.

As we have seen in other areas, the programs that compute and manipulate block models should be integrated with the general-purpose display and analysis routines. There is no need to duplicate these functions, as long as it is easy to retrieve data from the block model for input to other programs. This might require some special interface routines, since block models are usually stored in special formats to reduce size. The transformation from the special format to a more general one might be a function in the modeling package, or as a user option in the other programs.

After the reserve estimate is done, the block model is used in mine planning (as part of a feasibility study for the project). This is a modeling process in which operational constraints in extracting the ore are considered in conjunction with the value of the ore. For open-pit mines, various methods of pit optimization are used, to determine the portion of the orebody that can be mined profitably. For underground mining, interactive graphics programs are used to design the layout of ramps, shafts, and so on. These procedures are in the realm of development engineering, and not part of the exploration process.

Geostatistics

Because of the many uncertainties outlined above, ore reserves usually now are computed using geostatistics. This is especially true for deposits that will be mined by surface methods, since the representation of the deposit by a set of contiguous blocks closely approximates the way the ore will be extracted. Geostatistical methods are also used in selecting which mined blocks are ore, using more detailed assays acquired during the mining operations.

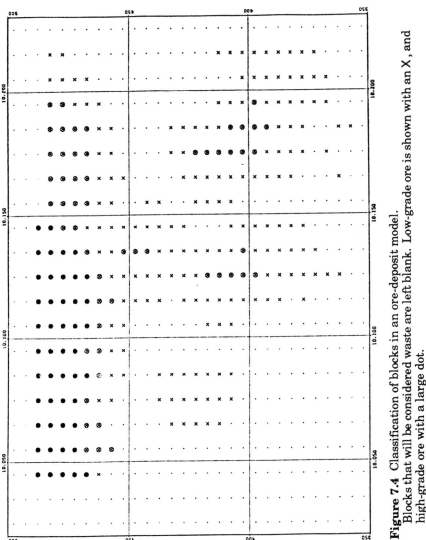

Figure 7.4 Classification of blocks in an ore-deposit model. Blocks that will be considered waste are left blank. Low-grade ore is shown with an X, and high-grade ore with a large dot.

Geostatistical gridding methods provide the best estimate of the ore grades within each block, in the sense that the block grades are derived by an error-minimization process. These methods are called kriging, with variations having qualified names such as indicator kriging. (This section assumes an understanding of the basic principles of geostatistics; see the Bibliography for more information.)

When ore reserves are defined by specific ore-zone boundaries, geostatistics can also be applied. Although most commonly applied to block averages or point estimates, kriging can be adapted to calculate the average value within an arbitrary volume. This is precisely what is needed in sectional reserves.

There are three main steps to a geostatistical model for an ore deposit. First the variogram (or set of variograms) that defines spatial variability of the data must be calculated. Next, a variogram model is developed, to express the observed form of the experimental variograms in a tractable mathematical form. Finally, the desired type of kriging calculation is run on all the blocks in the model.

There is no absolute right way to do the job. Although the superiority of geostatistics over alternate methods has been clearly demonstrated many times, there is no corresponding consensus on the variety of ways of doing geostatistics. Several different types of variogram models can be used, to be combined with at least as many types of kriging. Further complicating the problem, the process of defining a variogram model is highly subjective, depending on the experience and inclination of the analyst.

The variogram and kriging calculations can be very computer intensive, requiring programs that are highly tuned for computational efficiency. Much of the computer time is spent in searching for data, similar to nongeostatistical gridding. Some of the techniques to speed up gridding can be adapted for geostatistics. Similar limitations on the amount of data to use in estimation are also applied here. The main difference between kriging and other estimation procedures is that a matrix inversion is required for each value to be estimated. Thus, kriging programs must employ highly efficient matrix manipulation routines, as well as rapid searching.

The results of kriging are stored in the block model in exactly the same way as other estimates (i.e., there is nothing inherently different about the kriged value). The need to verify the results is also the same, as are the methods used. There is one difference, though: Kriging also produces an estimation variance for each block. It can be displayed (e.g., as another set of contours) to show areas where the estimates are most uncertain. The estimation variance is controlled primarily by the geometry of points used in estimating the block. As one might expect, it tends to be low where the data density is high, and higher where fewer data are available and the data distribution is highly nonuniform. If the behavior of the estimation variance is to be examined in detail, another set of maps may be required. Provided there is not too much clutter, it might also be shown as an additional variable on the maps of estimated grade. Techniques for creating such composite maps have been outlined in earlier chapters.

ESTIMATING RESERVES IN PETROLEUM EXPLORATION

The estimation of reserves for an oil or natural gas field is similar to mineral-reserve estimation. The emphasis on computing the volume of the field is stronger, since volume is the primary factor in judging the field's potential value. It is still necessary to determine the values of variables such as porosity throughout the field, although these values tend to vary more slowly than mineral grades, which means that is easier to interpolate values.

The volume estimate is derived from the interpreted structural model for the oil reservoir, in essentially the same way as a sectional ore reserve for a variable thickness vein. The final result of interpreting the seismic and well data is a series of structural maps, showing the elevations at the structural boundaries that form the top of the reservoir. The base of the reservoir might be another structure, or perhaps a planar surface at the oil–water contact. The areal limits may be defined by the intersection of the oil–water contact with the upper surface, but are likely to be controlled also by fault boundaries and other controls. The reserve is frequently measured within a land-ownership boundary. Any such spatial limits are digitized as part of data preparation for the volume calculation.

Volume computation is normally based on a gridded representation of the structural surface. Because of the large potential value of the oil contained in the reservoir, the grid must be a highly accurate representation of the surface. In the situation where several partners have interests in an oil field, the judgment of whether the grid is indeed correct can be contentious.

Although gridding relies on the assumption of continuity in the data, this is not always the case in structural mapping. Many oil fields are influenced strongly by faults, which displace segments of the original structure into disjointed fragments. The gridding procedure must be applied independently within each fragment. Since the displacements can be in any direction, determining the precise boundaries for each is a complex geometrical problem. When there are two surfaces to be generated, the fault boundaries must be interpolated carefully between them. In effect, the faults add many additional surfaces to the enclosing volume for the reservoir.

Rather than using a standard gridding program independently on each block and combining the results, the constraints imposed by the fault boundaries are usually considered in the gridding program. This allows a one-pass gridding procedure to be used. The operation of the program (from the user's point view as well as computationally) is necessarily more complex. In addition to the elevation data, a set of digitized fault traces must be supplied. Additional parameters may be needed to control how the fault traces will be used. For example, if the fault displacements are small, it may be reasonable to use all data for a first-pass gridding calculation, and then limit data within specific fault zones when computing the final grid. Alternately, a distance of influence of the fault may be specified. The option to use data across fault boundaries is sometimes an unavoidable compromise, since there may not be sufficient data strictly within the fault zone to allow an accurate grid-point estimation.

As in other three-dimensional mapping examples, it is difficult to verify the results of the gridding process. Once again, preparation of a series of maps comparing original data to the gridded model is needed. Values from the model may be interpolated at the well locations, so that differences from the observed values in the wells can be highlighted on the maps.

Geostatistical techniques for gridding are becoming more commonly used in oil exploration, although they are not yet as widespread in the mining industry. This is probably due to the nature of the data: Mineral grades are more erratic and unpredictable than structural surfaces. The uncertainty in estimation is thus higher in mining: The need for methods that quantify the uncertainty was a major stimulus of the development of geostatistics. Associated variables such as porosity and gas pressure may not be as well behaved as the structural surfaces, leading to greater uncertainty in estimation (and thus a greater incentive to use geostatistics).

Developing a gridded model of the oil-bearing structures is just the beginning. The volume of a structure and the amount of oil contained it in are the real objectives. As discussed above, the volume is derived by adding the volumes of all the grid cells that are within the reservoir. Unlike mining applications, where volume calculations to the nearest block are usually sufficient, in oil reserves the partial blocks must often be calculated with high accuracy. (A partial block occurs where a fault or lease boundary cuts through a grid cell.) This is a requirement in multi-owner situations, since the extraction of fluid in one place causes movement of fluids away from others. It is important to accurately measure the proportion of the oil in place within each ownership, before pumping starts.

The digitized fault or lease boundaries then are also input to the volumetrics routine. For each grid block, a test against all the boundaries is needed to determine whether the entire volume is included in the total. If not, the partial volume inside the boundary must be computed. With nonvertical faults, the calculation must be done as an intersection of a plane with a cube (or rectangular prism). For ownership boundaries (which generally extend vertically into the ground), it is a matter of computing the partial area. The thickness to assign to the partial block must also be considered: It is not sufficient to simply take the grid-point value. Depending on the gradient of the structure around the block, the value to use may be higher or lower. What is needed is an estimate of the thickness at the centroid of the partial block. As in all the other applications of estimation, there may be several methods available to do this interpolation.

To compute the actual amount of oil or gas, the porosity within the blocks also must be considered. The amount of water present is another factor, since it will occupy some of the pore space. Partial saturation is another possibility that will affect the amount of oil or gas. For natural gas, which is highly compressible, the pressure distribution is also a variable to be considered.

In developing the reservoir, the extraction process is modeled using block models of all of the above parameters. Reservoir simulation uses large computers to test different plans for the production phase of the oil field. Rates of pumping, injection of water or steam to maintain pressure, and many other techniques have a major effect on the amount of oil or gas that actually comes out of the ground. Typically this is much less than the amount in place (30% is a typical recovery rate), so the operating procedures can significantly change the economics of the field. This is of lesser importance in mining, especially in open-pit mines, where a greater percentage of the resource will be extracted. Like mine planning, these considerations are part of development engineering, which follows after the exploration phase is complete.

COAL RESERVES

Strip mining of coal is similar to open-pit mining of metals, but the size and nature of the operation requires some special treatment of reserve estimates. It also has some similarities to oil reserves, in that the emphasis is on the volume of the coal seam more than on the grade. Here grade means the unit heat content and amount of contaminants such as sulfur. This factor is important, but not the critical parameter in determining the amount of the economic resource.

The amount of coal in a seam is determined by a two-dimensional approach. Gridded models for the top and bottom of the seam are created, from which the variable thickness of the seam can be calculated. The seam thickness also might be gridded directly, although it is likely that the top and bottom surfaces will be needed in planning the stripping operation (e.g., for amount of overburden to be removed).

The scale of the operation is a source of problems. Strip mines might be several kilometers on a side, and the spacing of data is necessarily large (hundreds of meters between drillholes is typical). Although the boundaries of the seam tend to vary slowly, there is obviously a considerable degree of uncertainty in estimating the thickness between the drillholes. Geostatistical estimation is used to create grids of surfaces and thicknesses, as a result.

The volume calculation from the gridded data is essentially the same as for the oil reservoir case. The constraints of lease boundaries or faults may be present. The heat content and amount of contaminants may have to be considered in adding the blocks. Limits on these values are part of the contract with the buyers, so that some areas within the seam may be excluded from the reserve. Alternately there may be more than one class of reserve, each with a different price to be considered in economic evaluation.

WATER RESOURCES

The procedures for estimating the volume of water in an aquifer are similar to the oil reservoir case, with the additional complication that there is a variation with time. The total volume within the aquifer (based on its boundaries and porosity) is fixed, but from year to year, or at different times of the year, the amount of water held in the aquifer may be quite different. In addition to collecting data to define the aquifer, it is then necessary to have data on how the water content varies. The water resource will be a range, with the minimum level being of most concern in planning.

The water content may be based on direct observations from monitoring wells, although for new development this will not be adequate, since the time period of the data will be too short. In this situation, an estimate of how much water enters the aquifer from the surface will be required. Data on precipitation and runoff rates from point stations are used to estimate the total influx of water over the catchment area. This gridding procedure is based on sparse and erratic data in many cases. The estimation must be done for each time period available in the data. Geostatistics is applied to this type of estimation, as well.

The source of water also may be snow or ice fields in mountainous regions. The volume of the snow pack in spring indicates how much water will be available through the summer. This is conceptually a standard volume calculation. Given the difficulties of access to the snow-pack areas, the amount of data usually is limited. Rough topography means that the surfaces are complex, and thus more difficult to interpolate. The accuracy of these estimates is necessarily low, as a result. Remote-sensing methods are applied to this problem, as satellite images are the only practical way to track the changes in the snow cover through the year.

CHAPTER 8

Other Computer Applications

There are many other uses for computers in exploration than have been discussed to this point. This chapter is an overview of some of these applications. Some are closely related to the methods in previous chapters, while others are more general in nature, using techniques that are widespread in data processing (i.e., they are not as specific to the geosciences). As we have noted many times, the use of general-purpose programs in a variety of situations is more efficient that extensive customization of programs for a specific job. This philosophy will also be evident in this chapter.

The topics to follow are diverse and largely independent. They fall into two categories: use of specific types of software in exploration, and specialized tasks that are developing in the field. The Bibliography lists sources of further information.

PREPARATION OF REPORTS

The results of an exploration project are almost always summarized in a written report. In addition to descriptive text, reports contain graphical data (such as location maps, cross-sections, analytical displays) and tables (lists of laboratory measurements, ore reserves in various zones, summary statistics of different types of data, etc). With the current generation of word-processing software, all of this information can be prepared and combined using a computer.

Considerable training and experience is needed to create a complete report with a word-processing program, so it is generally not done by the analyst. However, it is not necessary for all of the document preparation needs to be left to a skilled operator, only the final composition. The text of the report is written by the analyst, typically using a personal computer. Depending on the analyst's computer expertise, and inclination to experiment, much of the final document layout may be done at this stage. With most word-processing programs, functions such as defining headings and page layout are relatively straightforward. Some people use a computer as an electronic notepad, writing fairly unstructured text that is later reorganized by an experienced operator.

For major projects, several people may contribute to the report; thus, several text files will be combined. They may be in somewhat different formats, reflecting the individual style of the authors. Using the same word-processing program and the same defaults for the document layout can reduce the effort in combining the pieces into a consistent whole.

The analyst also creates many of the graphical displays and tables for the report. Output from data-analysis programs must be saved in files that can be retrieved by the word-processing programs. For tables and some statistical summaries, these files are simply additional text, to be inserted in the report like any other text.

Integration of graphics is a more complicated step. Although it can be fully automated within some word-processing programs, it may be easier to simply leave appropriate spaces in the document for manual insertion of maps, and so forth. Graphics integration is currently a major area of software development, since it is the key to desktop publishing. Programs with this description are now available on personal computers. and are rapidly spreading the ability to create quality reports beyond the publishing industry.

Graphics integration requires a link between the word-processing software and the programs that create the plots. The plot instructions must be in a format acceptable to the word-processing software. Scaling to fit the page must be considered, usually by trial and error. Unless the document is prepared using a high-resolution graphics screen, the form may be checked only after printing the page.

SPREADSHEETS

Along with word-processing programs, spreadsheet programs were a main reason for the popularity of personal computers. They continue to be one of the most popular types of software, in terms of number of copies purchased. This is one area where the personal computer was a major innovation: Spreadsheet programs did not exist on larger computers prior to the advent of personal computers. Most other types of software developed first on larger machines, and were later adapted for use on smaller ones.

A spreadsheet program in its simplest form provides the ability to manipulate rows and columns of numbers, for example, monthly sales and expenses for a small business. Some of the columns may be text, as needed for descriptive purposes. Operations on rows (adding up sales for a month, perhaps) or by column (sales in a year by category) are used to define new rows and columns of data. These operations are defined by mathematical formulae, which can include complex forms, and logical tests (e.g., sums of one column when another exceeds some threshold). Columns that are defined by operations are automatically updated; that is, whenever one of the fields used in the formulae is changed, the result is changed.

Although the examples above describe simple business applications (which are the main use of spreadsheets), the nature of the data is secondary to the process. Many types of exploration data fall naturally into this row–column format. The extensive abilities of spreadsheets for entering and checking data, printing reports, sorting, and plotting simple graphs make them very useful in exploration, without even considering the advantages of the mathematical operations.

For example, geochemical analyses are presented in tabular format, with a sample identifier, and various columns of numbers representing the amounts of different elements. Additional descriptive data and sample location information may have to be added manually. A spreadsheet editing function allows new columns to be added more easily than with conventional text editors. The additional data may already be in another file (e.g., coordinates from a digitzer) which can be inserted as new columns. In other words, the spreadsheet can provide many of the merging functions needed in data preparation. Some of the data values may be reported as "trace" or below a detection limit, using a < symbol. These entries must be adjusted to true numeric form before calculating statistics. It may be necessary to calculate new variables that are combinations of the existing ones, for example, total concentration of heavy metals.

As we saw in earlier chapters, similar procedures are used in preparing many types of data for analysis. Rather than writing special programs for each type, it may be more efficient to spend the time gaining experience with a single flexible program such as a spreadsheet.

Spreadsheets are useful in testing a number of "what if" questions. Ore reserves are found by adding the contributions from all blocks in a block model that meet the current definition of ore. The test might be simple (a single metal grade above a cutoff), or complex (a combined measure of metals, in a particular type of rock, with maximum allowed contaminant levels). With the tabulated values from the block model, a spreadsheet can quickly test variations in these criteria, and produce a printed report. A variety of options for the form of the report are usually available.

Note, however, that a specialized program may still be needed, especially when using the results for detailed planning. The advantage of the spreadsheet is its adaptability to a variety of situations; it should not be expected to cover the in-depth analytical requirements.

The more traditional business applications of spreadsheets should not be forgotten. Running an exploration project involves accounting and project management as well as field work, mapping, and analysis. Tracking expenses, work progress, supplies, and so on can become a major job for the project manager (who also may be the chief data collector and analyst). Financial forecasting for variable costs, interest rates, and so on, can be easily implemented.

Although the use of a spreadsheet for these tasks is conceptually quite different from data analysis, much of the experience with the program is applicable to either situation. For example, the commands to print a report, methods for adding or correcting data, and defining calculations will be exactly the same. This is another reminder of the benefits from using general-purpose software for tasks that on the surface seem different.

REMOTE COMMUNICATIONS

Exchange of information between computer systems is a growing need in almost all businesses and scientific organizations. This task is simpler than it was even a few years ago, with readily available software and higher-speed communications hardware. The emphasis is now more on sending and receiving messages, rather than on connecting to a more powerful computer at a major center. Data may be transferred to a central location, but this is done more for archiving purposes than for using greater processing power.

Service companies in the exploration business, for example, geophysical survey crews and geochemical laboratories, routinely provide their customers with results in computer format. This procedure may involve physical transfers on diskettes or tapes, or electronic transfer via telephone/modem connection. The latter method requires a computer account to be set up, and the customer to learn some basics of operating the contractor's system. It can become a minor problem when dealing with a number of service companies, since each will have a different procedure for logging in, checking results, transferring data, and so on. The same procedure on the user's end should apply in all cases, which helps reduce the difficulty.

In addition to receiving data and sending data, remote computer links are useful for sending messages from field operations to a supporting office. Detailed lists of required equipment, operational schedules, and so on can be typed on a personal computer at the field site, and then transferred as a file after connecting to the support center. This reduces connection time, and provides clearer and more detailed information than a verbal conversation.

Connection to other computers is also a key to effective use of the many data bases that relate to the geosciences. Abstracts of scientific journals, published data on companies, financial trends, and many others may be searched through a variety of computer services. (More details on use of data bases will follow in the next section.) Depending on the location of the user and the data base, it is often possible to use shared packet networks for the computer link, instead of normal phone lines. These are generally less expensive, and provide data checking to minimize the chance of introducing errors in data transmission.

USE OF LARGE DATA BASES

Another trend in the geosciences is that government agencies that monitor activities are providing public access to much of the information they collect. In some industries, similar data bases are available through service companies. As noted above, access to the data might be via on-line connection, and remote use of the computer where the data base is maintained. Alternately, the agency that created the data base may choose to distribute it on tapes or diskettes. If the data base is dynamic (i.e., subject to frequent revisions and additions), the on-line method is preferable.

Examples may serve to illustrate some ways to use such information. First we will consider a mineral inventory data base, which is an index to all known mineral deposits or occurrences in a particular area. Basic data for each occurrence include the location, name(s), list of the metals/minerals present, and current status (exploration project, in production planning, operating mine, etc.). Additional information might include a geological description, types of exploration work done, ownership history, and literature references. In the most complete cases, full text of reports might also be stored in the data base. In the oil industry, data bases are available with records of all oil wells in particular areas. Information includes the total depth, geological formations penetrated, formations tested, and amount of hydrocarbons present.

The primary reason for creating such data bases is to save time in retrieving information that otherwise would be in several places. Having all the data in one place opens the door to other uses, such as selecting records based on particular characteristics. Many data base retrieval programs are used for these applications. If access is via the source computer, the user is not faced with a software choice. When purchased data are installed on the user's machine, software must be selected to provide the type of search and reporting abilities needed.

Full-text management systems are best for data bases containing extensive text (e.g., detailed geological descriptions), while conventional data bases may be more appropriate if numerical manipulation is required. Full-text systems create a cross-reference table of all words in the data base, without requiring special structuring of the data. Thus, memos, reports, and so forth can be stored in the data base in original form, and it is not necessary to build tables of key words for searching. Note that in computer publications listing software, full-text management is a distinct category from data base management, even though there are many similar features.

What type of information can be extracted from such files? This question is open-ended: The more different types of data present, the greater the combinations that might be used as search criteria. The simplest case probably is to select by name, that is, to find the information related to an occurrence with a specific name. Since mineral claims tend to have common names, this search might turn up many records.

In regional exploration, patterns of mineral distribution are useful guides. This is another simple search, using a list of names or codes to be matched. The selection can be an AND test (i.e., all entries in the list must be found), or an OR test (select if any entry is found). The primary output is a map showing the locations of all records from the search. Since maps are not normally an option in the data base search program, this process has two stages. The search program creates a file with coordinates, and other basic information such as names and status. It may be necessary to edit this file to make it compatible with the map-drawing program, although this usually can be done automatically. Plotting the map then follows the same procedures as plotting other types of data. The map may be a composite of data from the search and other sources. For example, in Figure 8.1, all silver occurrences for an area in British Columbia are shown as points on a symbolic map of silver in geochemical samples.

Another common type of data base is a record of historical activity in exploration, for example, oil wells drilled each year. Searching on the date, with output to a file for plotting a map, provides a view of changing

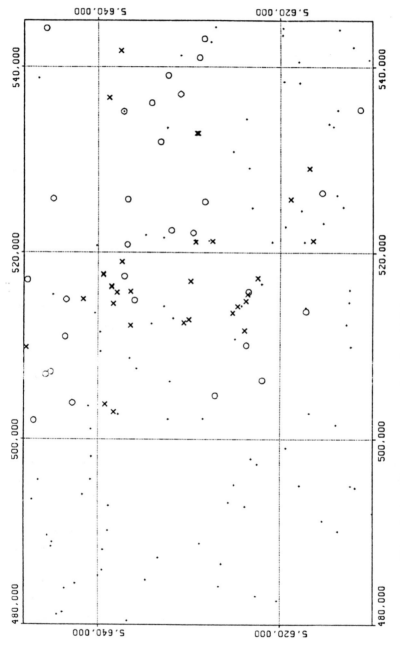

Figure 8.1 Map of mineral occurrences and geochemical data. Known mineral occurrences containing silver are marked by X. Geochemical sample locations are marked by a dot; those with anomalous silver content are circled.

patterns. In Figure 8.2, a series of maps showing mineral exploration activity in British Columbia illustrates how areas of heavy activity come and go with time. This type of display can point out areas that may have been overlooked for renewed exploration. The difference in the maps of the 1970s and 1980s reflects a switch in exploration objectives, from large porphyry copper targets to smaller gold deposits. In analyzing oil exploration, similar maps are used. Here there is an additional variable to consider: the depth of the well. Oil may occur in multiple formations at depth, so it is necessary to plot maps showing wells that penetrated particular formations. Once again, such maps may show relatively unexplored areas that would not be evident on a map simply showing all wells in the region.

An alternate form of display for occurrence data is shown in Figure 8.3. Density of occurrence is contoured, to give a quantitative measure of the differences between the areas with clusters of points in Figure 8.2B. Contouring has the usual assumption of spatial continuity: In this case, the gridding calculation is simply a count of all points found within the grid cell. The objective of contouring is to highlight trends connecting clusters, in hopes of finding promising areas for exploration. This procedure could be applied to any set of data extracted from the data base. As in the earlier example, the map might also show other data for the same area.

More complex search criteria may be applied and viewed in the same way. The best strategy for a complex search is to do it in stages. The number of records that match the criteria may quickly become small, so that it is easier to simply look at all of the information for these records than to continue searching (which likely will not further reduce the count, or reduce it to zero). For example, an exploration history file might be searched for all properties on which company X did drilling and radiometric surveys in 1981, and for which ore reserves were calculated. It may turn out that only three records show any work by company X in 1981, so there is not much point in carrying on past this point. The ease of combining searches is inherent in the software, and should be considered as an important function when evaluating programs.

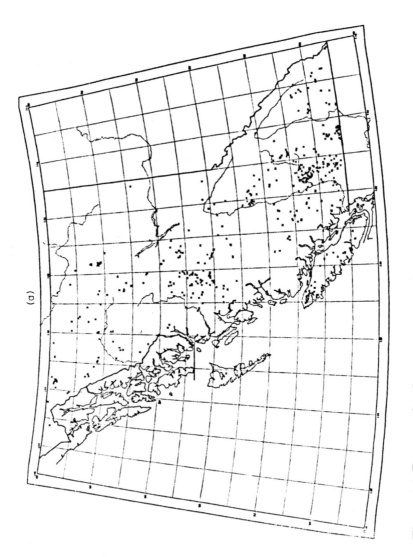

(a)

Figure 8.2 Patterns of exploration activity in British Columbia.
A: Mining claims worked in 1970.   B: Mining claims worked in 1985.

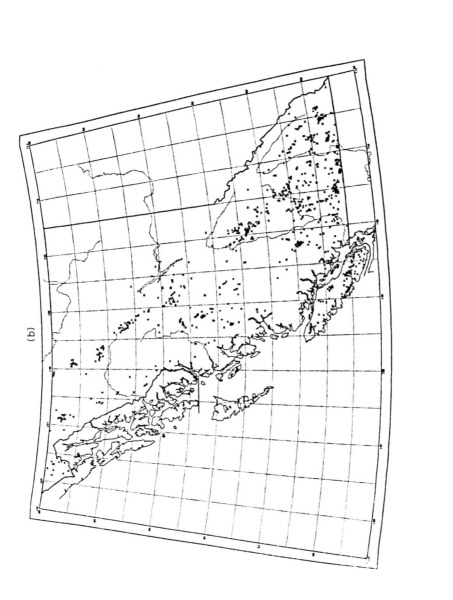

(b)

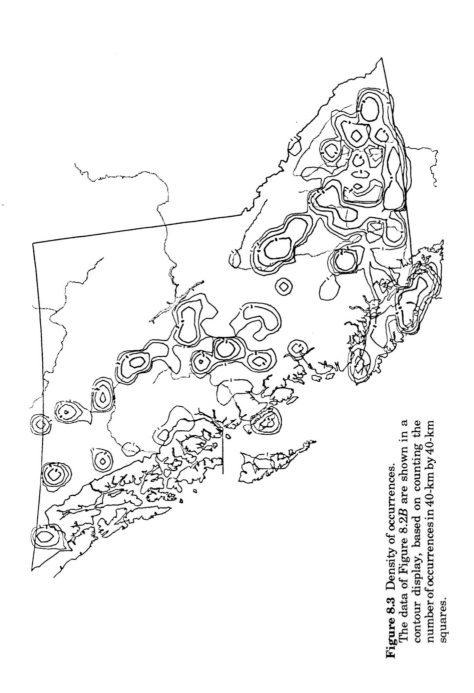

Figure 8.3 Density of occurrences. The data of Figure 8.2B are shown in a contour display, based on counting the number of occurrences in 40-km by 40-km squares.

GEOGRAPHIC INFORMATION SYSTEMS

The benefits of combining several types of data on a map have been mentioned several times. Most of the illustrations in earlier chapters did not include additional cartographic details, to concentrate on the display of the analytical data. The usefulness of a map is greatly increased when topographic and cultural features are shown along with the exploration data (e.g., Figs. 8.2 and 8.3). If these enhancements are omitted, the analyst has to use two (or more) maps to see location reference, check for correlations with other features, and so on. If three-dimensional information is available, the data may be viewed in true spatial perspective (Fig. 8.4).

A general-purpose mapping program can create combined displays, although usage can become cumbersome when many different types of data are involved. File storage and retrieval also becomes a problem, especially when several different areas are involved. Each will have its own files of topographic data, drainage patterns, cultural features, and so on, in addition to the exploration data.

The solution to these difficulties is software combining large data base management attributes, map projection and drawing, and efficient graphical display. Programs of this type are called geographic information systems (GIS). As the name implies, they were developed originally for cartographic applications such as land-use planning, but can also be applied in other fields such as exploration.

These systems demand considerable computing power. Therefore, they were not implemented on personal computers until the late 1980s. To produce cartographic quality maps interactively, high-speed and high-resolution terminals are needed. The fine detail for contour lines and similar features means large files are needed, and large, fast disk drives to store them. Hard copy will be in the form of large maps, so a large fast plotter is also an essential component.

Like other complex systems, the power and flexibility offered by a GIS does not come free. First of course is the expense of the hardware and software, which will be substantially higher than for simpler mapping systems. The need for specialized training and the time to become proficient with the system will also be greater. The initial setup of the system and all the desired data files is likely to take a long time.

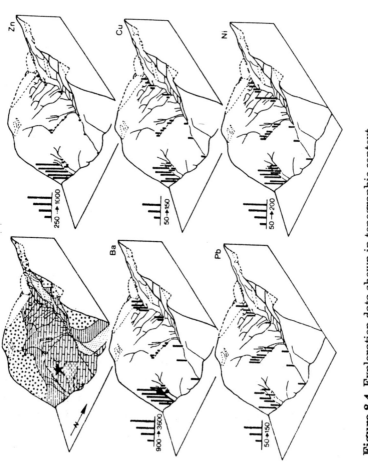

Figure 8.4 Exploration data shown in topographic context. (Reprinted from *Journal of Geochemical Exploration*, vol. 31, p. 124, 1989. "Geochemical stream sediment and overburden surveys of a zinc- and barium-mineralized area, Freuchen Land, central North Greenland," by U. H. Jakobsen.)

Once the data base for a project has been created, the GIS takes over the job of integrating new information as it is acquired. Whenever new maps are to be drawn, or new analyses are to be done, all of the previous data can be retrieved. Throughout the life of a major project, the vast amount of data quickly becomes uncontrollable without a sophisticated system or heavy investment of manpower.

In addition to keeping track of the graphical data, a GIS also provides ways to test spatial relationships. Various attributes are associated with the graphical data, which can also be used in analysis. Searching with a GIS is similar to data base searches, except that matching is based on common geographic area as well as on specific attributes of the data. For example, in a land-use application, one might want to find all agricultural areas above a given elevation, or where annual rainfall is low, or within a given distance of a major population center. In municipal mapping, a typical problem might be to locate all streets with water systems over a certain age, and display them along with such other utilities as power lines. This information might be stored in a variety of ways, for example, point locations, gridded values, or polygonal outlines. The software has corresponding functions to locate points within polygons, intersections of polygons, and so on.

Exploration applications of this type of spatial analysis are essentially unlimited. One of the major problems in interpreting multidisciplinary surveys is that the data are not taken at the same locations. With a GIS, corresponding zones can be matched up quickly, without necessarily relying on gridding techniques to bring them to a common reference. For example, regional geochemical and geophysical surveys could be subdivided according to geologic boundaries, to test the statistical relationships of the data within each zone. This type of analysis could be done with conventional programs, of course. The advantage of a GIS is that the software is designed to do these tasks quickly, and in an interactive graphics environment, so that the results are displayed immediately for visual interpretation.

Simple comparison of regions defined by polygons is often useful. This type of matching is much less likely to be available in conventional software. For example, find all claims within a given region, or that border the claims of another company. In planning exploration programs, it is useful to know how much of an area is covered by lakes or heavy vegetation.

Another type of analysis deals specifically with the topographic data. Using the data in a digital terrain model (DTM), the GIS software can determine the boundaries of drainage systems (used in analysis of geochemical surveys), estimate cut-and-fill volumes, and perform a variety of similar tasks. Like spatial analysis, it is not essential to use a GIS for these applications, but advisable to do so for efficiency.

The speed of a GIS largely depends on clever techniques for storing and displaying line-type features. Some of these developments may be useful outside a GIS, in other situations where large digitized files are used. Programs are available to scan digitized data to retain only the significant points. An alternate approach is to determine a mathematical description of each feature, using fractal geometry. Depending on the nature of the data, the effective number of points on the line may be reduced to a small fraction of the original value.

ARTIFICIAL INTELLIGENCE: EXPERT SYSTEMS

A long-time dream of computer system designers is a computer that can reason like a human. Most computer programs operate with precise rules and rigid procedures, unlike human reasoning, which is highly subjective, even in well-defined scientific disciplines. A completely general system of this type is far in the future, but for certain well-defined problems the objective can be partially met.

One type of program for artificial intelligence is an expert system, which attempts to mimic the reasoning process of an human expert in some field. (Other branches of artificial intelligence include robotics, speech recognition, machine vision, and natural language processing.) Applications are in areas where rigid mathematical procedures are not sufficient for solving a problem, but a considerable level of experience and intuition is needed. In the areas we have been discussing, this means primarily at the interpretation and modeling stage.

Expert systems have two major components. First is the knowledge base, which is simply the human experts' knowledge about the subject coded into a computer file. Second is the inference engine, which is a set of rules the expert would use to interpret new data. In addition to the reasoning program, software to aid in creating the knowledge base and the rules for the inference engine is included in a complete expert-system package.

To set up an expert system for any subjective task is not a trivial procedure. First, the experts who provide the information for the system must express their experience and interpretive methods in a logical form. Doing so may take some careful thinking and analysis, since some of the steps may become almost subconscious as experience grows. The information then is coded into the form required by the system. This step is fairly routine, since most expert systems include routines for entering data interactively. It may be time-consuming, since the volume of information is typically large. (If the knowledge base is small and the rules are simple, there is not much incentive to go to the trouble of building an expert system.)

Like the human expert, the computer system should be able to learn from experience. In other words, results of using the program are added to the knowledge base. Over time, the program should be able to do a more complete analysis of new situations. There are some implications for the computer system itself, in that the increased size of the knowledge base may increase the execution time, and eventually use up an excessive amount of disk space.

There are many benefits to be gained from creating an expert system. First, it provides a way of capturing at least some of the valuable knowledge of experts in the field, for use by others who are less experienced. In most cases, this knowledge is lost to the organization when the person retires or goes on to other jobs. The knowledge can be disseminated widely, simply by distributing copies of the system (the human can only be in one place at a time). As a training method, an expert system can be useful, as it will show the user what types of information are critical to the problem at hand, and demonstrate how they are interconnected. Even if the system does not provide definitive answers, these benefits can more than justify the cost of development.

The highly subjective nature of interpretation in exploration makes it difficult to create expert systems that are widely applicable. One of the early technical successes of expert systems was the PROSPECTOR project at Stanford University, which used a set of geologic model descriptions to locate mineral deposits. Although the trial project used a variety of geologic criteria, its models were restricted in type to the geologic environment for which it was designed (porphyry deposits in the North American Cordillera). The extensive work that would be required to generalize the system to different geologic situations has prevented wider use of this technique.

In petroleum exploration and development, expert systems have been used for well-log interpretation. This problem is less complex than identifying mineral deposits (using single wells, not three-dimensional bodies, and using similar types of data, rather than a variety of types). As a result, expert systems have achieved greater success in well logging. This illustrates a fairly common observation in the artificial intelligence world that suitable problems fall into a middle range of complexity. If they are too complex, it is not practical to organize all of the rules and knowledge into a computer system. As observed earlier, if they are too simple, there is little to be gained from setting up an expert system.

EXPLORATION DECISION MAKING

In all types of exploration choices must be made on the methods, level of detail, time spent, and so on. Inevitably, it must be decided whether the objectives of exploration have been met (i.e., whether to stop the project). In the broadest sense, all of the computer methods described in this book contribute to making these decisions. As used here, exploration decision making means using computers specifically to make judgments about the viability of projects, value of information, and similar economic parameters.

Exploration for resources is done in stages, with the results of each stage determining whether or not to proceed to the next stage. For example, a regional study may start with analysis of satellite images. If interesting geological features are found in the images, the next stage might be airborne geophysical surveys and regional geochemical sampling. If anomalies are found in these data, more detailed surveys may be done over these areas. This might comprise several stages, with ever-decreasing spacing of ground magnetics, seismic lines, soil samples, and so on. Based on this information, it may be decided to test the anomalies by drilling.

The decisions at each stage are seldom clear-cut. There are almost always some anomalies found: The question is whether they indicate a reasonable chance of success at the next stage. Further complicating the picture are the limited funds available: The amount of work is almost never enough to provide definitive answers. The objective of decision-making software is to quantify these problems in terms of probabilities of success.

For the earliest stages of exploration, the problem is to define the types of exploration to be done. The undiscovered bodies and structures are considered as targets to be detected using one or more of the available methods. Geologic models of the potential targets are an integral part of this planning. Estimates of potential size and configuration of the targets are derived from existing knowledge of the area, and by analogy to areas with similar geology. In addition, the probability of detecting such targets with various exploration methods must be estimated. These estimates then can be used in an optimizing program, which tests various combinations of surveys to find the one that gives the maximum chance of finding one of the model targets. The minimum size of a potential discovery that would be considered economic is also a factor.

When proceeding from the regional to a more local scale, it is desirable to reduce the area to be examined in more detail, by producing a map of favorability (i.e., to define the most likely areas for finding an ore deposit, oil field, etc.). This procedure requires more detail in the geologic model, to take into account existing data such as satellite imagery. Programs for computing the favorability measure have similarities to expert systems, as well as to conventional data analysis. The process is in effect finding anomalies in the available data, and ranking them according to heuristic rules on the significance of each type of anomaly. For example, a geochemical reading five times the background level might have to be ranked in comparison to a geophysical anomaly observed across three survey lines.

At later stages, the problem may be more to determine the amount of data needed to ensure a given chance of detection. The objective might be to have a 75 percent probability of finding any structure over 500 meters in length, for example. This is largely a matter of spacing of survey lines and stations, which will depend both on the expected size of the target, and on the region of influence of the survey method. Geochemical sampling is aimed at finding dispersion halos, which can be much larger than the source body. Geophysical responses can be measured beyond the boundaries of the body, to varying distances. Estimates of the size and magnitude of anomalies then are needed to derive the probability of detection.

At later stages, decisions become dominated by economic concerns. When sufficient data have been acquired to estimate an ore or oil reserve, monetary values can be assigned. Estimates of the probability of increasing the reserve are used to determine the value of further exploration. For example, if some boundaries of the target are still "open" (i.e., the position of the boundary cannot be determined from the available data), the cost of drilling some additional holes can be balanced against the potential increase in the reserve. At the point of deciding whether to develop the resource, there are many additional considerations that go beyond the exploration aspects. Future variations in commodity prices, interest rates, production rates, and many other unknowns must be estimated (usually as a range of possibilities). This type of analysis is done by financial specialists, and thus goes beyond our frame of reference.

Guidelines for Effective Computer Analysis

This book consists of a rapid and perhaps too-cursory view of many useful techniques. The main point to remember is that computer applications seldom need to be restricted to the field for which they were developed. Many of the techniques were developed for purposes much different from analysis of data by geoscientists. (Many multivariate techniques were first applied in the social sciences, for example.) Some of them started in specific geosciences (geostatistics, for example, for estimation of mineral reserves), but are now widely used in other fields.

For anyone who uses computers, an open outlook on all aspects of computer technology is invaluable. The only constant is change: Users of computers must strive to keep up-to-date with computer developments in general as well as their own specialty. This is admittedly a tall order, but one that seems unavoidable. The rapid development of computer systems, and their rapid decline into obsolescence, has been going on ever since the beginnings of the computer age. It seems highly unlikely that this situation will change soon.

What goes along with the constant change in computer systems is the need for the user to experiment. No matter how detailed a description of what a computer system can do, testing its applicability to real problems is the only way to determine whether it is useful. This often means using a system in ways different from its intended purpose.

Experimentation does not only apply to new systems either: A computer system that has been in place for a year or two has not necessarily had all its uses tried, much less perfected.

The first thing to remember about computer data analysis is to be flexible. If the book has made this point clearly, it has met my objective of helping geoscientists to use computers effectively.

Throughout this book, a variety of different computer methods have been described. While the emphasis was on specific tasks related to specific types of data, a great deal of common ground was noted. This section will summarize the common features that should be incorporated into computer programs, to make it easy to use several different programs in conjunction with each other.

FLEXIBILITY

In almost every chapter, the need to try a variety of different techniques (or variations on the same technique) was noted. While it is impossible to anticipate all possible ways a program might be used when it is being written, it is essential to define all the important parameters, and to avoid making any assumptions about them.

EASE OF USE

Ease of use is essential to allow the user to take advantage of flexibility. Unfortunately it places conflicting demands on the programmer: The more options the user has, the more difficult it becomes to use the program. The solution is for the program to make reasonable estimates for the parameters as much as possible, so that acceptable results can be obtained without having to exhaustively consider all the options.

For complex programs that require many parameters, it is essential to allow the user to save a set of parameters for later use. Interactive parameter setting and testing is also needed, but it should be possible to bypass these steps and use parameters from a file. This combined approach allows the user to tune up the parameters for a particular program (say plotting a map with several types of data and a detailed legend), and then use these parameters for another run (say plotting the same style of map for an adjacent area).

COMMON FORMS OF INPUT AND OUTPUT

No single program can do everything, or even a majority of the tasks that require a computer. Several different programs will often be used in sequence to get a particular job done. This approach is preferable to trying to provide a great range of options in a single program, which usually results in an unmanageable programming problem. It also avoids redundancy, which inevitably results from trying to make self-contained systems.

To make the transition between programs simple, they must use a common data format. Thus, the output from one program (say a modeling program for gravity data, or a multivariate statistical analysis package) can be used as input to another (e.g., a map drawing program).

The user interface (i.e., the method and style of supplying instructions to the program) should be as similar as possible in all programs. This means consistent naming of parameters and commands to allow experience with one program to be applicable to others.

GENERALITY

Since a variety of data types are used, the software should not assume any particular type. A flexible data structure is necessary, and the terminology in the user interface must be general, that is, without reference to a specific type of data. For example, a program written for analysis of geochemical data might use terms like *element* and *mineral fraction*, which would be confusing when applied to geophysical data.

ABILITY TO RUN CONCURRENT TASKS

Another aspect of using a collection of programs is that it is often useful to run more than one program at the same time. If the operating system does not allow this, it may at least be possible to suspend operation of one program to use another. For example, in using a contouring package, a histogram of data values can be very useful in choosing contour levels. If the histogram is not prepared ahead of time, interrupting the contour program to run a histogram can save time (as opposed to stopping completely and having to start over from the beginning after preparing the histogram).

Multiple tasks are even more effective in a windowing system, which allows the results from several programs to be visible at the same time. Following the example above, the histogram could be displayed in one window, the contouring parameters in another, and the contour display in a third. Windows may still be used in a single-task operating system, with the restriction that only one task is active at any one time. The display of each window remains visible.

These capabilities are developed only as a result of close interaction between the developers and users of the software. Users inevitably find new ways of using software, beyond the original concept of the programmer. In doing so, they may find bugs, or unneccessarily complicated procedures that can be fixed easily. In an organization with its own programs and programmers, it is important to maintain good communication between the two groups. For those who buy software, this is not always possible, but it should be considered as a factor in choosing software, as discussed in Appendix B.

Selecting a Complex Software System

Although there are off-the-shelf programs for many of the tasks of exploration data analysis, the need for individual options inevitably becomes important. This is especially true for the more complex jobs such as reserve estimation, as discussed in Chapter 7. The ability of a program to adapt to individual requirements becomes a major factor in selecting which program to buy.

This appendix is adapted from a presentation at the *Workshop on Computerized Mine-Planning Systems,* held by the Canadian Institute of Mining and Metallurgy in Toronto in May 1987. It reviews some of the criteria that need to be considered when selecting any large software package. The terminology is oriented toward choosing a system for complex data evaluation, such as ore reserves and mine planning. The principles are quite general, however, and should be applied to any evaluation of large complex programs for scientific data analysis.

Choosing a computer system for data analysis and modeling is not a simple task. There are many methods for doing these jobs, and a great range of parameters involved in making decisions based on the results of using the programs. There is a corresponding variety of systems available, with a wide range of capabilities (and price).

There are also many special characteristics of any organization doing exploration and resource development (e.g., the type of operation, location, and level of staff computer experience). Existing computer applications might also have an effect on how the new applications will be incorporated.

Because of these variations, it is impossible to develop hard and fast rules for selecting a computer system. The choice depends to a large degree on judgments that cannot be reduced to hard analytical terms. It is possible to narrow down the range to be considered, however. This discussion provides guidelines on how this may be done, along with suggestions on how to effectively rate competing systems. The final choice of a computer system must come from within an organization, however: It is not reasonable to expect an outsider (even a highly experienced one) to understand the local variations in the necessary detail.

THE NEED TO COMPUTERIZE

Since traditional methods for data evaluation are manual, a common problem is to decide when (and if) it is appropriate to convert to a computerized system. Unfortunately, this is one of the most difficult questions, since there are seldom simple ways to measure the benefits obtained from installing a system. In some cases, there may be existing expenses that can be compared to estimates of what they might be after computerization.

How much is being spent on computer services?

It may be that the decision is not whether to switch from manual to computer methods, but whether to do the work outside or within the organization. This case is probably the easiest to test on the basis of costs, since the amount being spent on outside services will be replaced by the internal cost of buying, installing, and running the system. It is important to do a careful review of the internal costs, since it is easy to overlook the expense of training your staff, and having the additional task of running the computer added to the office workload.

How often are new evaluations needed?

One of the major advantages of using a computer is the time saved on repetitive tasks. The first time a computer does a job, it often takes just as long as using a manual method (because of the time spent in setting up the proper instructions to the machine). Later updates generally require only a small fraction of the time, since there is very little setup time. Manual updates usually need nearly the time of the original job, however. As an example, consider the drafting of a set of cross-sections or plans. Redrawing by computer simply requires the new information to be added to a computer file, and the same programs are run as before. Whenever the volume of new data is sufficient, the manual approach demands that the sections be completely redrafted.

To judge whether it is worth going to a computer system, it is then necessary to estimate how much of the work will be routine revisions, and how much will be one-time jobs. It is worth noting that updates will almost certainly be done more frequently when a computer is available, so that there may not be a cost saving, but rather more productivity for the same expense.

Are staff available to do computer work internally?

In the case of a new office being opened, this question may be simpler than when a computer system is being added to an existing operation. For example, the choice may be between two people for manual drafting, and one operator and a computer-aided design (CAD) system.

When the computer is placed in an existing operation, it must be determined whether current staff can take on the job of running the computer, or whether new people with computer skills are required. This decision is dependent in part on whether the computer is to replace certain manual tasks, or to provide functions not currently available. The degree of training must be considered, along with the expected workload; for example, is running the computer to be a part-time addition to current duties?

It is important to note that computers do not generally lead to a reduction in staff. Justifying a computer system on the basis of cutting personnel is likely to prove to be a mistake. The general pattern of computer usage is that the introduction of computers leads to many new jobs being done, rather than simply making existing jobs faster. More often than not, the end result is more staff, not less.

Does the computer provide new capabilities ?

The last point above is related to another major reason for using computers: The machine can do jobs that cannot be done any other way. In estimating reserves for potential mines or oil fields, for example, it is necessary to deal with vast amounts of data, or to employ complex calculations such as geostatistics, or both. This area can also be difficult to put on a cost–benefit basis: What is the value of the information derived by the computer? In the case of geostatistics, the question may be: What is the value of the reduced uncertainty of the result when compared to traditional methods? There are no general rules to apply here: The decision is highly subjective, and often depends mainly on the philosophy of the organization.

BASIC PARAMETERS FOR SYSTEM SELECTION

Assuming that a decision has been made to get a computer system (or new software), the first step is to define the basic criteria that the system must meet. The objective is to narrow the range of choices, so that a detailed evaluation is required for only a few systems.

How much can you afford?

One of the major distinguishing features of complex software systems is price. Although it is not practical to set a firm cost range prior to checking out the available systems, it should not be hard to set a rough limit. It may immediately eliminate many possible choices. This cutoff should not be too inflexible, however; detailed evaluation may show that the economics of installing the system are different than expected. It is also advisable *not* to eliminate very inexpensive systems: Price is not always an indication of capability. Even though you have decided you can spend $20,000, you may find an acceptable system for $5,000.

What is the system intended to do?

The key is whether the applications are varied, or focused on a single problem. A computer system dedicated to developing operating plans for one mine will not necessarily work well for developing ore reserves and trial mining plans for a number of prospective mines. In the latter case, much more flexibility is required to accommodate the different parameters of each new project.

Who will operate the system?

It makes a considerable difference whether the system is to be run by a computer professional or geologists/engineers. A computer person is more likely to tolerate cryptic notations, complex file systems, and the like than someone whose primary expertise is not in computers. The nonexpert will expect more on-line help features, easily understood prompts and error messages, and other such niceties.

It also matters whether the system is used very frequently, or only occasionally. With everyday use, the operator (whether or not a computer professional) should have the ability to shortcut some of the input procedures to streamline the use of the computer. When used only at periodic times, more tutorial/instructional help is desirable, since the procedures for using the system never get totally committed to memory.

Should two or three dimensions be emphasized?

The nature of data evaluation may be primarily two-dimensional (maps of surfaces or thin structures) or three-dimensional (block models of volumes). When the system is for an existing operation, this choice is obvious. When it is needed for evaluations of potential new operations, both situations may arise. Since it may not be possible to find one system that handles both cases adequately, it may be necessary to consider buying two different software packages. Ideally they should operate with the same type of procedures. In practice, it is unlikely that this goal can be achieved, since the buyer does not have complete freedom to modify the programs.

In mining, for example, an underground planning system requires more sophisticated computer graphics than an open-pit system. Underground layouts need to be done in three dimensions, while open pits usually can be reduced to a series of two-dimensional cases. Special software must be included to do the graphics for the underground case, but it is possible to use general-purpose mapping programs to handle many of the needs of open-pit planning.

What type of output is required?

The main purpose of the system may be to provide numeric data for incorporation into existing procedures for plotting maps, statistical displays, and so forth. More commonly the system will be expected to provide a variety of graphical output, either through its own plotting software, or by creating files suitable for use in a CAD system. A complete list of forms of output should be made, to use in the initial phase of evaluation.

Is this computer system entirely new?

If the organization does not currently have a computer, the system requires both hardware and software. The choice is more difficult, since questions of computer capacity, expandability, and so on must be considered. Also, it is likely that no one in the organization will have experience in such matters.

The system may be simply a software purchase to run on existing equipment. This method is not always the best, however, as the most suitable software may not run on the existing hardware. In addition, it is important to consider the workload on the computer, and whether it can handle the extra tasks. It may well be advantageous to buy hardware specifically for the new application, even if other hardware is already in place.

Is the system to be dedicated to the application?

Note that the system is almost certain to be used for other tasks as well as those defined in selection. Look for general purpose hardware, and make sure software does not place a barrier in the way of running other programs, even if the hardware is designated as only for the specific job. The additional uses of the machine may also come from running other programs to support the software specific to the job. For example, spreadsheet programs are often used to do financial calculations on mining models.

If the intent is to share a computer for evaluations and other tasks, it is obviously even more important to demand flexibility. It is also important to define what information (if any) needs to be passed between tasks, which may limit the types of data bases used by each program.

FIND OUT WHAT IS AVAILABLE

The next step is get a list of available systems, and some basic information on each one. Scan the standard journals in your field and visit trade shows at annual meetings of related organizations. Appendix C lists some of these sources. With the rapid spread of computers in the geosciences, it is likely that you have personal contacts who have recently developed their own shopping list.

After getting a list, contact the vendors for basic information on each system. The type of material in standard sales brochures should be sufficient, although approximate price quotations will usually be given separately.

EVALUATION OF VARIOUS SYSTEMS: PHASE I

Now compare the basic requirements and the information on the candidate systems to come up with a short list for detailed evaluation.

Select software before worrying about hardware.

This is the first rule for choosing any new computer system, not just data-analysis systems. Software is almost entirely responsible for how effective the system will be, as it controls the ease of use, flexibility, options for output, and so on.

Check essential criteria and your unique aspects.

The basic requirements are developed as shown above. It may be useful to rate these in importance, and decide if some can be changed, since it is possible that no system will meet all of the needs. If your operation has unique or unusual factors of major importance, these should be at the top of the list. For example, a mine may have a complex polymetallic ore, with several metals to be considered in ore reserves and economics.

Determine the adaptability of the system.

The need for flexibility arises in almost all computer applications. It is less important in the single-operation case, but still is an important consideration, as over time methods of entering data and displaying results tend to change. When the system is applied to multiple projects, it is possible that each will have its own form of input and unique type of output. The easier it is to adapt the system to these variations, the more productive it will be.

Find out if the source code is available.

This is another aspect that is of more concern in the situation where different types of evaluations are needed (rather than in a single operation). The advantage of having the source code is that custom modifications not anticipated by the developers of the system can be implemented by the user. When fixing bugs, it is usually faster to incorporate corrections and recompile than to get an updated version from the vendor.

Offsetting these benefits are the increased cost (in some cases several times more), and the need to have considerable computer experience to incorporate local changes. There is also a potential for creating unique problems that are no longer supported by a vendor's warranty or maintenance agreement.

Some programs include a programmers' toolbox, which gives access to the source code. Compiled versions of the basic functions are provided, along with programming instructions, so that the user can write programs without having to learn complete details of how the system works. The system functions can be invoked with a one-line statement in the user program. In an image-analysis system, for example, the toolbox would include functions for selecting subareas of the image, storing derived quantities, displaying the results, and so on.

Evaluate the level of support and training.

A major concern with any computer system is that it stays in operation. For remote locations, be sure that there will be good access to service when required. More than one service agreement may be necessary for the various hardware components and the software. The cost of service (or a service contract) must also be considered in budgeting for the system.

Depending on the complexity of the system, it may take a long time to become fully operational, and even longer to use it efficiently. The initial warranty should include a reasonable period of time when the user can consult with the vendor on special problems or difficulties. If possible, an in-depth hands-on training program should be given to the operators of the system. This additional expense can be overlooked in budgeting.

Find out how many versions are operational.

The more systems that have been sold, the better the chance that inconsistencies, bugs, and so on have been worked out. While commercial success is not always a guarantee that the system will work well in your operation, it is often the only indicator of a product's quality. This is not to say that you should automatically avoid a new vendor's system, but you will have to approach it with more caution.

Contact other users for comments.

A natural extension of the above is to get opinions on various systems from people who are currently using them. Try to get more than one for each system if possible, and aim for operations that are similar to yours.

Be aware of hardware considerations.

Very little has been said about selecting hardware, primarily because of the dominant role of software in the effectiveness of the system. There is another factor that contributes to the secondary role of hardware. Many software packages are tailored to a given hardware configuration, so that choosing software may provide the list of hardware by default. If the system runs on more than one configuration, local considerations often narrow down the choice (e.g., only one of the options may be readily available and serviceable in your area).

WHAT TO LOOK FOR IN SOFTWARE

Since the characteristics of the software will control the decision, some general principles of rating software are useful.

Is it flexible in format of input/output?

This point has already been made above, but cannot be overemphasized. No matter how rigidly data formats are defined, variations are bound to develop. If the system cannot easily adapt to the variations, a good deal of time and effort can be spent in modifying data bases to be acceptable to a program.

Is it easy to use and learn?

Ease of use is largely a matter of consistency of procedures. The methods for accessing data bases, defining parameters, and producing output are very similar (if not identical) in all phases of the system. The learning period can be greatly reduced if all software adheres to this principle. As noted earlier, consistency is not always possible when use of the computer system involves running several programs from different sources.

Part of the learning curve and ease of use depends on personal preferences of the operator. Procedures that seem clear and simple to one person may be obscure and frustrating to another.

Is it independent from particular hardware/software?

The rapid technological advances in computer systems, and inevitably increasing demands placed on them, mean that the current hardware will likely be replaced within a relatively short period of time. If the software has been successful, you will want to move it onto another system with a minimum of disruption to your operation.

Other components of a complete system are also likely to be changed. If possible, the software should be easily adaptable to take advantage of such changes (e.g., adding a more sophisticated graphics terminal or plotter). Changes can include system software, if the planning program relies on other packages (e.g., improved data base or sorting routines).

What are the limits on size of data bases and so on?

The need to avoid size limitations is similar to format flexibility, and stems back to the same basic need to be adaptable to new situations. If the system has built-in limits (which almost all have), make sure the size of your current data base (and expected expansions) are far below the maximum. It is also worth checking into what is needed to expand the limits, if that becomes necessary.

How long does it take to run?

Most of the concern with software is with external aspects. These are the most important to the operator who has to supply instructions to the machine quickly and accurately. How long the machine then takes to do the job is also a consideration, especially if the nature of the work calls for tight deadlines. Simple hardware modifications (such as floating point chips) often can make a dramatic difference, and should be considered if the initial time estimates appear unacceptable.

Is there a background mode?

Even if a typical job will take a long time, it may not be a problem. The computer operating system may allow the option of running more than one task concurrently, so that the long jobs can be put into background. Thus, the operator can do something else while the job is running. The programs to be run in background must be structured to accept all parameters at the beginning, before starting extensive calculations.

It may also be possible to chain tasks to run in sequence, even if the computer does not support multiple concurrent jobs. Long jobs can be set up to run overnight, without having the operator present. The programs must get parameters at the outset, or allow them to be defined in a file that can be prebuilt.

EVALUATION PHASE II: TEST PROGRAMS ON SHORT-LIST

Presuming that the initial review has identified a few systems that might meet your needs, the next stage is to test each system. It may not always be possible to obtain trial versions, in which case you may have to rely on interviews with other users to supply this information.

Perform a sufficiently long hands-on trial.

Since the critical aspect of each system is how well it runs in your operation, the test should be done by someone who will be at least involved in operating it later. It should be run on your own data, and exercise enough of the options to judge if it can do all that is expected of it.

Determine how easy it is to run.

This instruction is self-explanatory, but not always simple to apply in rating two systems. Unless one is clearly superior in all phases, it may be that one is easier to apply in some areas, but more difficult in others. A ranking of relative importance of each phase might be needed to resolve ambiguities.

Evaluate how easily an experienced user can streamline procedures.

This factor is hard to judge on first trial, but important on any heavily used program. After doing the same job a few times, it can be annoying to continue to use a detailed prompting session designed for a novice. A few demonstrations from experienced users should be sufficient to make this judgment.

Determine how well the system fits in with other software.

No single software package will provide all the functions needed. Text editors, data-base management programs, plotting programs, and others may be used in conjunction with the evaluation software. If a variety of programs are used in an integrated fashion, it is essential that data files can be freely interchanged between them. As noted earlier, it is desirable to have a similar form of operation, as well.

Test the speed of execution.

Although the criteria above are likely to define your choice, it is important to ensure sufficient speed of execution to get the required jobs done. Elaborate timing tests are not essential at this point, particularly since they cannot duplicate the eventual running environment and tuning that may be possible to increase speed. If more than one system meets all the above requirements, the results of timing tests may indicate that one has an edge.

MAKING A CHOICE

Having gone through the tests of the candidate systems, it should be clear which of them can meet your requirements.

Can you afford the system that is best on technical grounds?

The answer is probably a simple yes or no, unless the price is not too far over the limit. In this case, a second look at the cost justification may be in order.

If not, is a second choice still acceptable?

Once again, answer yes or no, then look at the price again. Variations in pricing structure might also have to be considered. The system may be sold as a series of semi-independent modules, some of which you might be able to do without to reduce the cost.

If no system is acceptable, what options are there for customizing?

Making a customized version can be dangerous, as it will not receive the same level of support as the standard product. It is also likely to be very expensive to have the vendor make custom changes. An organization with expert computer staff may not have a problem here, if they are willing to pay the premium to get source code for the package.

An alternate approach if no system appears suitable is to reexamine the criteria for selecting the system. After having looked at a variety of systems, and talked to other users, it is possible that the parameters you set initially will prove to be unrealistic.

GUIDELINES/RULES

This section is partly a summary of the points already made, and partly general rules that apply to several phases of the process.

Consider software first.

The fundamental role of software has been noted above. Another point is not to try too hard to adapt to existing equipment. You may save money in the long term by buying hardware specifically for the job at hand, rather than going to great lengths to make the software run on another system.

NEVER believe the salesman.

This is not to imply that they are all crooks: The problem is often that they are so experienced with the system that they don't recognize problems a new user will have. In addition, no one can judge how other people will react to the assumptions a program makes, its style of input, and so forth.

Demonstrations by an experienced user may look deceptively easy: You cannot tell how many hours of setup were involved, or how much experience with the system is needed to develop the same level of skill. In testing a program, be sure you can do the required tasks yourself.

Do not buy until the system has successfully worked with your data.

This consideration is related to the need for flexible input and output. Your data structure may be something the system has never handled before.

Try to avoid multivendor systems.

Unless you have staff experienced in running various types of hardware, get all equipment from one dealer and have it all under one maintenance agreement. When multiple vendors are involved, and one component of the system breaks down, the usual answer is "it's the other guy's problem" (which means it's really YOUR problem).

Remember the most important criteria: flexibility, flexibility, flexibility!

This point has already been made several times but bears repeating. It is impossible to anticipate all of the system's uses.

Keep in mind that no one system can do everything.

Integration with other programs is much easier than providing the same capability in one package. Be leery of systems that claim to be able to do everything. This might even be true, but it may mean learning how to use new programs to do jobs that you are doing already. It recalls also the need for passing data between various software packages.

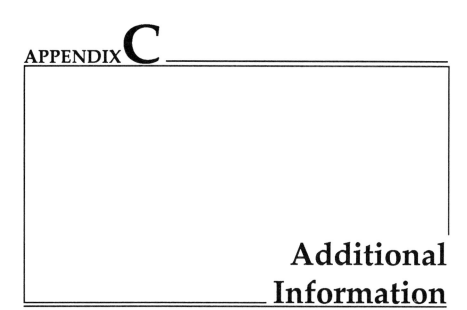

APPENDIX **C**

Additional Information

The rapid pace of development in computer technology makes it difficult to keep up-to-date. Whether you are looking for a system to buy, or interested in trying new techniques using an existing system, it is necessary to dedicate some time to this task.

In addition to the books and papers noted in the bibliography, information on using computers in the geosciences is available from several organizations. Many technical societies have branches specifically concerned with computer applications. Their annual meetings often have trade shows, which include exhibits of the latest hardware and software. A vast amount of software is available for a nominal charge from some of the organizations noted below.

ORGANIZATIONS

Computer Oriented Geological Society (COGS)
P.O. Box 1317
Denver, CO 80201-1317 USA

The monthly *COGSletter* is very useful for keeping up to date on meetings, new products, and so on. It includes user reviews of products and application case histories. COGS also runs an electronic bulletin board, and publishes an annual software directory, with summary information on a variety of systems for oil exploration, mine planning, and other earth science applications. It is of particular interest to people using small-scale systems (i.e., those based on personal computers). A set of programs for many geological applications is available to COGS members, for about the price of copying the diskettes.

Geological Information Group
Geological Society of London
c/o Dr. W. T. Sowerbutts
Department of Geology
The University of Manchester
Manchester, MK13 9PL United Kingdom

Similar to COGS in scope and functions provided to members.

Society of Exploration Geophysicists
P.O. Box 702740
Tulsa, OK 74170-2740 USA

European Association of Exploration Geophysicists
Wassenaarseweg 22
2596 CH, The Hague, The Netherlands

These organizations are the primary source of information on geophysical methods, particularly as applied to oil exploration. Their publications (*Geophysics* and *Geophysical Prospecting*, respectively) often cover computer applications. Many hardware and software companies have exhibits at their annual meetings. Many countries have similar societies, affiliated with the above.

Association of Exploration Geochemists
P.O. Box 523
Rexdale, Ontario M9W 5L4 Canada

This organization concentrates on applied geochemistry, particular in mineral exploration. Methods for multivariate analysis and display are often included in its *Journal of Geochemical Exploration.*

American Society of Photogrammetry and Remote Sensing
210 Little Falls Street
Falls Church, VA 22046 USA

In addition to the named fields, ASPRS and its journal (*Photogrammetric Engineering and Remote Sensing*) provide a forum for developments in geographic information systems.

United States Geological Survey
Books and Open-File Report Section
P.O. Box 25425, Federal Center
Denver, CO 80225 USA

Many of the programs used by USGS scientists are available, at very nominal costs. These usually include source code in standard languages such as FORTRAN. Most run on personal computers, but can be adapted to other machines. Applications include statistical analysis, geostatistics, gridding, and many others.

A separate branch of the USGS (US Geodata) sells digital elevation data, digitized coastlines, and other cartographic files, primarily for the United States. They also have software for cartographic applications. Contact:

National Cartographic Information Center
U.S. Geological Survey
507 National Center
Reston, VA 22092 USA

The geological survey departments of other countries also have developed many computer applications, which are available to the other geoscientists. For example, the Geological Survey of Canada was a pioneer in applying airborne geophysical methods and stream-sediment geochemistry to regional mapping. Like the USGS, they publish many of their computer algorithms at a nominal cost and are a primary source of exploration data.

National Technical Information Service
U.S. Department of Commerce
5285 Port Royal Road
Springfield, VA 22161 USA

NTIS publishes abstracts of reports in many scientific fields, for most U.S. government agencies, and projects funded by them. Related reports from other countries are also listed. Full reports can be purchased. Computer software is often included. Of particular interest is the *Natural Resources and Earth Sciences* newsletter.

The above list is selective. Many other geoscience organizations are active in publizing computer methods. *Geotimes* (the newsmagazine of the American Geological Institute) publishes a lengthy directory of organizations each year (e.g., Oct. 1989 issue).

SPECIAL REFERENCES

N. M. S. Rock, 1988, *Numerical Geology: A Source Guide, Glossary, and Selective Bibliography to Geological Uses of Computers and Statistics*, vol. 18 of Lecture Notes in Earth Sciences, Springer-Verlag, Berlin, London, New York, 427 p.

The subtitle says it all: It is an indispensable reference and source book. It contains an exhaustive (and exhausting for the author!) list of over 2,000 references. Each section has a summary list of references that covers basic theory, applications, computer programs, and so on. Sources of software are listed.

E. Hyatt, 1988, *Keyguide to Information Sources in Remote Sensing*, Mansell Publishing, London, New York, 274 p.

Similar to *Numerical Geology*, but on a more restricted subject. Includes information on publications, remote-sensing data sources, patents, conferences and other subjects.

MEETINGS

Applications of Computers in the Mineral Industry (APCOM)

A series of technical meetings covering all aspects of using computers in mining. Proceedings are published in book form and are a good source of state-of-the-art reviews. They are held about every eighteen months, alternating between opposite sides of the Atlantic Ocean (e.g., London, 1984; Pennsylvania, 1986; Johannesburg, 1987; Las Vegas, 1989; Berlin, 1990).

International Conferences on Geoscience Information (GEOINFO)

Held every four years, to review trends and new developments in creating data bases for the earth sciences (London, 1978; Golden, CO, 1982; Adelaide, 1986; Ottawa, 1990).

Bibliography

JOURNALS

Computers & Geosciences (Pergamon Press). Articles on many different applications of computers in all branches of earth science. Many papers include full listings of programs.

GEOBYTE (American Association of Petroleum Geologists). Computer applications for the oil exploration and development.

Mathematical Geology (Plenum Press). Research papers on statistical methods in the earth sciences. A primary source of information for new developments in geostatistics.

Photogrammetric Engineering and Remote Sensing (American Society for Photogrammetric Engineering and Remote Sensing). Research developments in remote sensing and geographic information systems, plus reviews of new products, publications, and meetings.

The journals of many technical societies have frequent papers describing computer methods.

BOOKS

General References/Collections

Agterberg, F. P., 1974, *Geomathematics: Mathematical Background and Geoscience Applications*, Elsevier, Amsterdam, 596p.

Chung, C.F., A.G. Fabbri, and R. Sinding-Larsen, eds., 1988, *Quantitative Analysis of Mineral and Energy Resources*, ATO ASI Series C, vol. 223, D. Reidel, Dordrecht, Holland, 738p. (Proceedings of the NATO ASI meeting at Lucca in 1986.)

Davis, J. C., 1986, *Statistics and Data Analysis in Geology*, 2nd ed., Wiley, New York, 646p. and one diskette. A standard reference, with emphasis on petroleum applications.

Garland, G. D., ed., 1989, *Proceedings of Exploration 87*, Ontario Geological Survey, Special Volume 3, 914p. (from the Third Decennial International Conference, Toronto September 1987). Applications of geochemistry and geophysics in mineral, groundwater, and geothermal exploration.

Green, W. R., 1985, *Computer-Aided Data Analysis: A Practical Guide*, Wiley, New York, 268p. Many aspects of using computers effectively and what to look for in software are covered. The book also reviews many of the basic principles of data analysis and computer graphics.

Hanley, J. T., and D. F. Merriam, eds., 1986, *Microcomputer Applications in Geology*, Pergamon, London, 258p.

Koch, G.S., and R. F. Link, 1970–1971, *Statistical Analysis of Geological Data*, Wiley, New York. (Reprinted in one volume by Dover Books, New York, 1980.) Also a standard, with applications in mining exploration.

Rock, N. M. S., 1988, *Numerical Geology: A source guide, glossary, and selective bibliography to geological uses of computers and statistics*, vol. 18 of Lecture Notes in Earth Sciences Springer-Verlag, Berlin, 427p. Indispensable for anyone using computers in the earth sciences. Topics include all facets of statistics, time series, analysis of orientation data, spatial analysis, classification, and many others.

Shore, J., 1985, *The Sachertorte Algorithm; and Other Antidotes to Computer Anxiety*, Viking, New York 270 p. (also in a Penguin reprint). A popular-style look at general principles of computers and how to use them.

Books on Various Subtopics

As noted throughout the book, there is a large degree of overlap between many of the exploration and computer processing methods. The groupings below reflect the main topic of the books: These should not be considered as exclusive.

Petroleum Exploration and Development

Davidson, M. J., ed., 1986, *Unconventional Methods in Exploration for Petroleum and Natural Gas*, Southern Methodist University Press, Dallas, 350p.

Jones, T. A., D. E. Hamilton, and C. R. Johnson, 1986, *Contouring Geologic Surfaces with the Computer*, Van Nostrand Reinhold, New York, 314p.

Robinson, J. E., 1982, *Computer Applications in Petroleum Geology*, Hutchinson Ross, Stroudsburg, PA, 164p.

Mineral Exploration

Edwards, R., and K. Atkinson, 1986, *Ore Deposit Geology and Its Influence on Mineral Exploration*, Chapman and Hall, London, 466 p.

Peters, W. C., 1987, *Exploration and Mining Geology*, 2nd ed., Wiley, New York, 685 p. A good practical guide to all the procedures and problems of mining exploration, including geophysical and geochemical surveys, and drilling.

Raleigh, C. B., ed., 1985, *Observation of the Continental Crust through Drilling*, Springer-Verlag, Berlin, 364p. A collection of papers on advances in drilling technology, for deep crustal studies, but also with regard to resource exploration

Hydrology

Brassington, R., 1988, *Field Hydrogeology*, Geological Society of London, London, 175 p.

Haan, C. T., 1977, *Statistical Methods in Hydrology*, Iowa State University Press, Ames, 378p.

The following are collections of useful papers on methods for mapping and predicting water resources. All are published by UNESCO (Paris):

Hydrological Maps, 1977, 204p.
Groundwater Models, 1982, 235p.
Groundwater in Hard Rocks, 1984, 228p.

Statistical and Graphical Data Analysis

Chambers, J. M., W. S. Cleveland, B. Kleiner, and P. A. Tukey, 1983, *Graphical Methods for Data Analysis*, Duxbury Press, Boston.

Cleveland, W. S., 1985, *The Elements of Graphing Data*, Wadsworth, Monterery, California, 323p.

Lewis, P., 1977, *Maps and Statistics*, Methuen, London, 318 p.

Mather, P. M., 1976, *Computational Methods of Multivariate Analysis in Physical Geography*, Wiley, London, 532p.

Pandit, S. M., and S. M. Wu, 1983, *Time Series and Systems Analysis with Applications*, Wiley, New York, 586p.

Size, W. B., 1987, *Use and Abuse of Statistical Methods in the Earth Sciences*, Oxford University Press, Oxford, 169p.

Tukey, J. W., 1977, *Exploratory Data Analysis*, Addison-Wesley, Reading, MA, 688 p.

Remote Sensing and Image Analysis

Fabbri, A. G., 1984, *Image Processing of Geological Data*, Van Nostrand Reinhold, New York, 244p.

Lillesand, T. M., and R.W. Kiefer, 1987, *Remote Sensing and Image Interpretation*, 2nd ed., Wiley, New York, 721p.

Mather, P. M., 1987, *Computer Processing of Remotely Sensed Images*, Wiley, Chichester, 352p.

Sabins, F. F., 1987, *Remote Sensing: Principles and Interpretation*, 2nd ed., W. H. Freeman, New York, 449p.

Szekielda, K.-H., 1988, *Satellite Monitoring of the Earth*, Wiley, New York, 326p.

Geostatistics and Ore Reserves

Clark, I., 1979, *Practical Geostatistics*, Applied Science Publishers, London, 129p.

David, M., 1977, *Geostatistical Ore Reserve Estimation*, Elsevier, Amsterdam, 364p.

David, M., 1988, *Handbook of Applied Advanced Geostatistical Ore Reserve Estimation*, Elsevier, Amsterdam, 232p.

Hohn, M. E., 1988, *Geostatistics and Petroleum Geology*, Van Nostrand Reinhold, New York, 264p.

Journel, A. G., and C.J. Huibregts, 1978, *Mining Geostatistics*, Academic Press, New York, 600p.

Matheron, G., and M. Armstrong, eds., 1987, *Geostatistical Case Studies*, D. Reidel, Dordrecht, Boston, 248p.

Geophysical Methods

Dobrin, M., and Savit, C., 1988, *Introduction to Geophysical Prospecting*, 4th ed., McGraw Hill, New York, 867p.

Robinson, E. S., and Coruh, C., 1988, *Basic Exploration Geophysics*, Wiley, New York, 562p.

Ulriksen, C. P. F., 1982, *Application of Impulse Radar to Civil Engineering*, Lund University, Sweden, 180p.

Yilmaz, O., 1987, *Seismic Data Processing*, Society of Exploration Geophysicists, Tulsa, OK, 536p.

Geochemical Methods

Howarth, R. J., ed., 1983, *Statistics and Data Analysis in Geochemical Prospecting*, vol.2 of *Handbook of Exploration Geochemistry*, Elsevier, Amsterdam, 438p.

Levinson, A. A., 1980, *Introduction to Exploration Geochemistry*, Applied Publishing, Wilmette, IL, 924p.

Thornton, I., ed., 1983, *Applied Environmental Geochemistry*, Academic Press, London, 501p.

Thornton, I., and R. Howarth, eds., 1986, *Applied Geochemistry in the 1980s*, Graham and Trotman, London, 347p.

Exploration Decision Making

DeGeoffroy, J. G., and T.K. Wignall, 1985, *Designing Optimal Strategies for Mineral Exploration*, Plenum Press, New York, 364p.

Harris, D. P., 1984, *Mineral Resources Appraisal*, Oxford University Press, 445p.

Harris, D. P., 1989, *Exploration Decisions: Economic Analysis and Modeling*, Wiley, New York.

Weiss, A., ed., 1989, *Proceedings of 21st APCOM*, Society of Mining Engineers, Littleton, Colorado. Pages 103–178 contain seven papers on various aspects of uncertainty, risk assessment, and decision making.

Wignall, T. K., and J.G. DeGeoffroy, 1987, *Statistical Models for Optimizing Mineral Exploration*, Plenum Press, New York, 432p.

Expert Systems and Artificial Intelligence

Harmon, P., R. Maus, and W. Morrissey, 1988, *Expert Systems: Tools and Applications*, Wiley, New York, 289p.

Rauch-Hindin, W. B., 1986, *Artificial Intelligence in Business, Science, and Industry*, 2 vol., Prentice-Hall, Englewood Cliffs, NJ, 331p. and 348p.

Waterman, D. A., 1986, *A Guide to Expert Systems*, Addison-Wesley, Reading, MA, 419p.

Other

Gardiner, V., and D.J. Unwin, eds., 1985, *Thematic Mapping Using Microcomputers*, special issue of *Computers & Geosciences*, vol. 11, no. 3.

Morrison, J. L., 1971, *Method-Induced Error in Isarithmic Mapping*, Tech. Mo,. CA-5, American Congress on Surveying and Mapping, 76p.

Shelley, E. P., ed., 1986, *Proceedings of the Third International Conference on Geoscience Information*, Australian Mineral Foundation, Adelaide.

Snyder, J. P., 1982, Map Projections Used by the U.S. Geological Survey, *GS Bulletin 1532*, U.S. Government Printing Office, Washington, D.C.

Spradley, L. H. 1985, *Surveying and Navigation for Geophysical Exploration*, IHRDC, Boston, MA, 289p.

Unwin, D., ed., 1989, *Fractals and the Geosciences*, special issue of *Computers & Geosciences*, vol. 15, no. 2.

PAPERS FROM TECHNICAL JOURNALS

This list is highly selective, primarily of papers that review particular problems in data processing and analysis.

Agterberg, F. P., 1989, Computer programs for mineral exploration, *Science*, vol. 245, pp. 76–81.

Bonham-Carter, G. F., F. P. Agterberg, and D. F. Wright, 1988, Integration of geological datasets for gold exploration in Nova Scotia, *Photogrammetric Engineering and Remote Sensing*, vol. 54, no. 11, pp. 1585–1592.

Cameron, K. L., Cameron, D. D, and P. C. Kelleher, 1988, Producing geological illustrations using PC-based computer-aided drafting, *Computers & Geosciences*, vol. 14, no. 3, pp. 291–297.

Clark, I., 1986, The art of cross validation in geostatistical applications, *Proceedings of 19th APCOM*, R.V. Ramani, ed., Society of Mining Engineers.

Eckstein, B. A., 1989, Evaluation of spline and weighted average interpolation algorithms, *Computers & Geosciences*, vol. 15, no.1, pp. 79–94. Considers some problems in gridding (see also the report by Morrison).

Grunsky, E. C., and Agterberg, F. P, 1988, Spatial and multivariate analysis of geochemical data from metavolcanic rocks in the Ben Nevis, Ontario, *Mathematical Geology*, vol. 20, no. 7, pp. 825–861.

Hatton, L., 1989, Computer science for geophysicists. Part IX: The reliability of software, geophysical and otherwise, *First Break*, vol. 7, no. 4, pp. 124–132. Part of a continuing series that provides valuable insights on the problems in writing and using complex software packages.

Jones, C. B., 1989, Cartographic name placement with Prolog, *IEEE Computer Graphics & Applications*, vol. 9, no. 9, pp. 36–47. Discusses the problem of text overposting, and an expert system approach to a solution.

Jones, T. A., 1989, The three faces of geological contouring, *Mathematical Geology*, vol. 21, no. 3, pp. 271–283.

Kowalik, W. S., and W. E. Glenn, Image processing of aeromagnetic data and integration with Landsat images for improved structural interpretation, *Geophysics*, vol. 52, no. 7, pp. 875–884.

Reddy, R. K. T., and Koch, G. S., 1988, A generalized model for evaluating area-potential in a mineral exploration program, *Mathematical Geology*, vol. 20, no. 3, pp. 227–241.

Rock, N. M. S., 1988, Summary statistics in geochemistry: A study of the performance of robust estimates, *Mathematical Geology*, vol. 20, no. 3, pp. 243–275.

Watson, D. F., and G. M. Philip, 1989, Measures of variablility for geological data, *Mathematical Geology*, vol. 21, no. 2, pp. 233–254.

PUBLISHED SOFTWARE

See also the journals noted above, and the book *Numerical Geology*. Appendix C lists other sources.

Books with Multiple Programs

Koch, G.S., 1987, *Exploration-Geochemical Data Analysis with the IBM PC*, Van Nostrand Reinhold, New York, 179p. and two diskettes. A good training tool: includes basis statistical analysis programs and test datasets.

Merriam, D.F., ed., 1989, *Statistical Methods for Resource Appraisal*, special issue of *Computers & Geosciences*, vol. 15, no. 4. Programs for some of the contributions at the NATO ASI meeting at Lucca in 1986.

Press, W. H., B. P. Flannery, S. A. Teukolsky, and W. T. Vetterling, 1987, *Numerical Recipes: The Art of Scientific Computing*, Cambridge University Press, London, 818p. (See also *Numerical Recipes* in C, same authors and publisher, 1988, 735p.) Source code for many fundamental algorithms used in data analysis (FORTRAN and Pascal in the first book; C in the second). Indispensable for anyone writing software. Programs can also be obtained on diskettes.

Wellerman, P. F., and D. C. Hoaglin, 1981, *Applications, Basics, and Computing of Exploratory Data Analysis*, Duxbury Press, Boston.

Specific Applications

As noted above, *Computers & Geosciences* is a major source of source code. Many books also provide listings of programs, or have supplementary publications or diskettes with programs (e.g., Davis, Koch, David, Mather, 1976).

Dettori, G., and B. Falcidieno, 1982, An algorithm for selecting main points on a line, *Computers & Geosciences*, vol. 8, no. 1, pp. 3–10. A useful routine for reducing size of digitized files.

Guth, P. L., E. K. Ressler, and T. S. Bacastow, 1987, Microcomputer program for manipulating large digital terrain models, *Computers & Geosciences*, vol. 13, no. 3, pp. 209–213.

Larkin, B. J., 1988, A FORTRAN 77 program to calculate areas of intersection between a set of grid blocks and polygons, *Computers & Geosciences*, vol. 14, no. 1, pp. 1–14.

Martz, L. W., and E. de Jong, 1988, Catch: a FORTRAN program for measuring catchment area, from digital elevation models, *Computers & Geosciences*, vol. 14, no. 5, pp. 627–640.

Rock, N. M. S., 1987, Robust: An interactive FORTRAN-77 package for exploratory data analysis, *Computers & Geosciences*, vol. 13, no. 5, pp. 463–494. Many different types of statistical analysis and display, using both parametric and non-parametric methods.

Salomon, K. W., 1978, An efficient point-in-polygon algorithm, *Computers & Geosciences*, vol. 4, pp. 173–178. Fundamental to many data selection problems.

Wessel, P., 1989, Xover: A cross-over error detector for track data, *Computers & Geosciences*, vol. 15, no. 3, pp. 333–346. Adjustment of line misties.

S.R. Yates, 1987, Contour: A FORTRAN algorithm for two-dimensional high-quality contouring, *Computers & Geosciences*, vol. 13, no. 1, pp. 61–76.

Index